S 156
K3457 1995.
AGE 6996

BCC/UCF LIBRARY, COCOA, FL 32922-6598

UCF — GIFT

DATE DUE	

Statistical Problem Solving in Quality Engineering

Other McGraw-Hill Quality Books fo Interest

BOXWELL • *Benchmarking for Competitive Advantage*
CARRUBBA • *Product Assurance Principles: Integrating Design and Quality Assurance*
CROSBY • *Let's Talk Quality*
CROSBY • *Quality Is Free*
CROSBY • *Quality without Tears*
FEIGENBAUM • *Total Quality Control, 3d ed., revised (Fortieth Anniversary Edition)*
GEVIRTZ • *Developing New Products with TDM*
GRANT, LEAVENWORTH • *Statistical Quality Control*
HRADESKY • *Total Quality Management Handbook*
HRADESKY • *Productivity and Quality Improvement*
IRESON, COOMBS • *Handbook of Reliability Engineering and Management*
JOHNSON • *ISO:9000*
JURAN, GRYNA • *Juran's Quality Control Handbook*
JURAN, GRYNA • *Quality Planning and Analysis*
MENON • *TQM in New Product Manufacturing*
MILLS • *The Quality Audit*
OTT, SCHILLING • *Process Quality Control*
PAKEIA • *Assurance Technologies*
ROSS • *Taguchi Techniques for Quality Engineering*
SAYLOR • *TQM Field Manual*
SLATER • *Integrated Process Management: A Quality Model*
SOIN • *Total Quality Control Essentials*
TAJIRI, GOTO • *TPM Implementation*
TAYLOR • *Optimization and Variation Reduction in Quality*
VANI • *The McGraw-Hill Certified Quality Engineer Examination Guide*

Statistical Problem Solving in Quality Engineering

Thomas J. Kazmierski

McGraw-Hill, Inc.
New York San Francisco Washington, D.C. Auckland Bogotá
Caracas Lisbon London Madrid Mexico City Milan
Montreal New Delhi San Juan Singapore
Sydney Tokyo Toronto

Library of Congress Cataloging-in-Publication Data

Kazmierski, Thomas J.
  Statistical problem solving in quality engineering / Thomas J. Kazmierski.
      p.   cm.
  Includes index.
  ISBN 0-07-034048-X
  1. Quality control—Statistical methods. 2. Problem solving—Statistical methods. I. Title.
  TS156.K3457   1995
  658.5'62'015195—dc20   95-12017
                                                                        CIP

Copyright © 1995 by McGraw-Hill, Inc. All rights reserved. Printed in the United States of America. Except as permitted under the United States Copyright Act of 1976, no part of this publication may be reproduced or distributed in any form or by any means, or stored in a data base or retrieval system, without the prior written permission of the publisher.

1 2 3 4 5 6 7 8 9 0   DOC/DOC   9 0 0 9 8 7 6 5

ISBN 0-07-034048-X

The sponsoring editor for this book was Harold B. Crawford, the editing supervisor was David E. Fogarty, and the production supervisor was Pamela A. Pelton. This book was set in Century Schoolbook by Ron Painter of McGraw-Hill's Professional Book Group composition unit.

Printed and bound by R. R. Donnelley & Sons Company.

This book is printed on recycled, acid-free paper containing a minimum of 50% recycled de-inked fiber.

Information contained in this work has been obtained by McGraw-Hill, Inc., from sources believed to be reliable. However, neither McGraw-Hill nor its authors guarantee the accuracy or completeness of any information published herein, and neither McGraw-Hill nor its authors shall be responsible for any errors, omissions, or damages arising out of use of this information. This work is published with the understanding that McGraw-Hill and its authors are supplying information but are not attempting to render engineering or other professional services. If such services are required, the assistance of an appropriate professional should be sought.

To my wife, Mary Ann

Contents

Preface   xi

Chapter 1. Introduction to Statistical Problem Solving   1

    1.1 Factors Affecting Global Competitiveness   2
    1.2 Elements of Total Quality Management   2
    1.3 Management Levels versus Tools Used   4
    1.4 Seven Stages of Quality Evolution   5
    1.5 Proactive (Prevention) versus Reactive (Detection) Style of Management   8
    1.6 Costs of Poor Quality   9
    1.7 Process Improvement Translates to Quality Improvement   12
    1.8 Common Pitfalls Encountered During Process Improvement   14
    1.9 Target Value Principle   15
    1.10 Summary   18

Chapter 2. Pareto Analysis and Brainstorming Techniques   19

    2.1 Styles of Management   19
    2.2 Pareto Analysis   20
    2.3 Data Analysis Must Yield Actionable Information   23
    2.4 Expanded Thinking   27
    2.5 Matrix Model   30
    2.6 The How-Why Diagram   33
    2.7 Gantt Chart   41
    2.8 Contingency Planning   42
    2.9 Summary   43

Chapter 3. Introduction to Variation and Statistics   45

    3.1 Variable Measurements   45
    3.2 Methods of Evaluating Variable Data   46
    3.3 Measures of Central Tendency   48
    3.4 Measures of Dispersion   49
    3.5 Z Score   52

viii   Contents

3.6	Histograms	55
3.7	Run Charts	60
3.8	Using Normal Probability Paper to Evaluate a Nonnormal Distribution	66
3.9	Taguchi Loss Function	67
3.10	Summary	76
3.11	Practice Problems	77

Chapter 4. Measurement System Analysis   81

4.1	Importance of Measurement System Analysis	81
4.2	Definitions	82
4.3	Steps in Preparing for a Measurement Systems Analysis	83
4.4	Measurement System Analysis—Long Method	83
4.5	Gage Correlation	85
4.6	Measurement System Analysis for Attribute Characteristics (Visual Inspection)	88

Chapter 5. Shewhart Control Charts   95

5.1	Variables Chart	96
5.2	Control versus Capability	97
5.3	Identifying Opportunities for Improvement	101
5.4	Preplanning Checklist	102
5.5	Case Study	103
5.6	Recalculation of Control Limits	110
5.7	Sampling Guidelines for Control Charts	114
5.8	Sampling Schemes	115
5.9	Sampling for Continuous or Batch Processes	123
5.10	Individual and Moving Range Chart	124
5.11	Moving Average and Moving Range Chart	128
5.12	Modified-Limits Chart	131
5.13	Multistream Control Chart	134
5.14	Control Charts for Low-Volume Applications	137
5.15	Summary	140

Chapter 6. Capability   143

6.1	Principle of Capability	143
6.2	The C_p Index	143
6.3	The C_{pk} Index	147
6.4	Larger-Is-Better Characteristics	151
6.5	Smaller-Is-Better Characteristics	152
6.6	Using Normal Probability Paper to Calculate Capability	155
6.7	Summary	156
6.8	Problems	157

Chapter 7. Variables Control Charts, Multivary Charting, and Precontrol Charts   159

7.1	Recording Process Changes on Control Charts	159

7.2	Visual Pattern Analysis of Control Charts	161
7.3	Summarizing Continuous Improvement	165
7.4	Median and Range Control Charts	167
7.5	Case Study: Foundry	171
7.6	Multivary Charting	180
7.7	Precontrol	185
7.8	Summary	189

Chapter 8. Attribute Control Charts — 191

8.1	How Attribute Control Charts Fit into the Process Improvement Effort	191
8.2	Types of Attribute Control Charts	192
8.3	Considerations Prior to Use of Attribute Control Charts	193
8.4	A p Chart with Constant Sample Size	194
8.5	A p Chart with Variable Sample Size	199
8.6	An np Chart with Constant Sample Size	204
8.7	A c Chart for Nonconformities	205
8.8	Summary	211

Chapter 9. Check Sheets — 213

9.1	Constructing a Check Sheet	213
9.2	Defect Check Sheet by Frequency and by Cost	214
9.3	Defect Location Check Sheet	216
9.4	Attribute Check Sheet	218
9.5	Variable Check Sheet	220
9.6	Truckline Delivery Performance Example	222
9.7	Summary	225

Chapter 10. Scatter Plots — 227

10.1	Example	228
10.2	Calculating the Best-Fit Line	230
10.3	Coefficient of Determination	231
10.4	Summary	232

Chapter 11. Design of Experiment — 235

11.1	Introduction	235
11.2	Case Study: Compression Strength	238
11.3	Data Transformation and Analysis	241
11.4	Case Study: Burst Strength	243
11.5	Summary	246

Index 249

Preface

High quality should be thought of as high levels of customer satisfaction, low product-development costs, short product-development times, high levels of productivity, with very few manufacturing, assembly, test, or field problems. This definition also applies to the service industries and the administrative operations of an organization. Market surveys should identify customers' wants and requirements. These wants must be translated into products or services that give high levels of customer satisfaction in a short period of time at low cost.

Throughout my career as an educator, engineer, and consultant I became very much aware of a need for people to be aware of the full range of techniques that should be used during the journey leading to process improvement. Some of these techniques are statistical in nature: histograms, control charts, capability studies, and designed experiments. Some other techniques are not statistical techniques: brainstorming techniques, how-why diagrams, and Pareto analysis.

The book is organized in the order that closely follows many process improvement efforts.

1. Senior management team determines list of long range measurable objectives. (Pareto analysis, brainstorming)
2. An action plan is developed to facilitate accomplishing the objectives. (how-why diagram)
3. Resources are allocated for the tasks that must be done and a completion schedule is made. (Gantt chart)
4. Data is gathered, summarized, and analyzed and action is taken when needed. (check sheet, control chart, capability analysis, scatter plot, designed experiment)
5. Unsatisfactory processes must be improved. Once the improvement is verified, the benefits must be measured and then communicated throughout the company. (brainstorming, designed experiments, process flow revision, cost of quality report)

Throughout the book the term *process* is used. A process could be a welding process or a process could be the process of conducting a market survey. New product development is a series of processes. Process improvement translates into: improvements in quality, customer satisfaction, cost reductions, shortened time to market, reduced field failures. Process improvement is the end result of many long hours in the trenches searching for the right combination of elements of the process that will yield the desired results and then the implementation of the new process.

This book stresses that it is best not to be so preoccupied with the capability index of a process and the theoretical proportion out of specification, but rather to scrutinize the data to verify that the process is in a state of statistical control.

Acknowledgments

I would like to thank Richard Miesowicz and Bhagwan (Ben) Ramnani for reviewing and critiquing the manuscript of this book. Thanks also to Dr. Jim Kowalick and Mr. Davis Bothe for their insights into the robust design techniques and methods for determining capability for the nonnormally distributed processes described in this book. I sincerely appreciate the guidance given to me by Patrick H. Norausky, Vice-President of Total Quality, Dresser-Rand Company. Lastly, I would like to acknowledge the past support given to me by the late Dr. Ben Vineyard of Pittsburg State University. Dr. Vineyard was instrumental in helping thousands of educators and engineers begin their careers.

Thomas J. Kazmierski

Chapter 1

Introduction to Statistical Problem Solving

This chapter will explain many ideas and principles that are the foundation of successfully using the technical tools of *statistical problem solving* (SPS). It is important that the reader feel comfortable with these ideas and principles, or else there could be difficulty in the application, analysis, and willingness to take action when the SPS charts suggest action is needed.

There are many different terms used in the language of quality today. *Statistical process control* (SPC) is probably the most common term. Many people would define SPC as using a control chart to help improve a process or a quality characteristic for a product.

The main deficiency of this definition is that the person may not be aware of many other simple statistical tools besides the control chart. Some of those tools are Pareto analysis, brainstorming techniques (cause and effect), measurement system analysis, histograms, run charts, check sheets, scatter plots, design of experiment, and so on.

A common term in the 1950s and 1960s was *statistical quality control* (SQC). In retrospect, much of the effort focused on improving the quality of the product. This is the desired goal, but many people failed to understand that the process that yields the product must be improved—then and only then will the quality of the product improve. The primary tool used in SQC efforts was the control chart.

In this book, the term *statistical problem solving* is used frequently. SPS is like a toolbox full of tools. Different tools should be used at different stages during the process improvement effort. The important point is that readers need to be very aware of a wide variety of statistical tools that can be used depending upon the situation. Usually

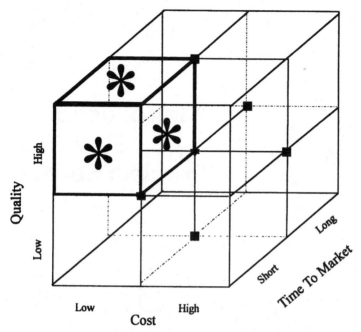
Figure 1.1 Factors affecting global competitiveness.

there is not any one right answer as to what tool should be used in certain situations.

1.1 Factors Affecting Global Competitiveness

The definition of quality in the past 10 to 20 years has broadened to encompass topics that were originally thought not to be related to quality. For a company to be competitive in today's markets, management must wisely balance the *quality/cost/time-to-market* relationship to meet or exceed customer requirements. Figure 1.1 shows a $2 \times 2 \times 2$ three-dimensional matrix where eight possible combinations are shown concerning quality, cost, and time to market. Clearly the desired combination is high quality, low cost, and short time to market.

The challenge that management faces is to transform the culture, technological skills, and daily work efforts to achieve the long-range objectives of the company. Optimizing the combination of quality, cost, and time-to-market factors will give companies a greater probability of being competitive in the future world economy.

1.2 Elements of Total Quality Management

Using the technical tools of statistical problem solving is only one of the

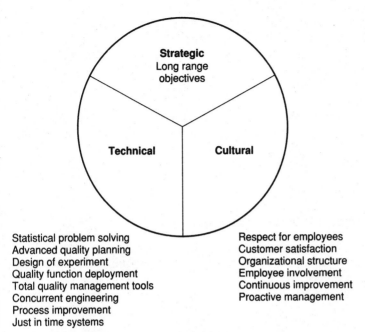

Figure 1.2 Elements of a comprehensive TQM system. (*Reprinted with permission of the American Supplier Institute, Inc.*)

three major elements of the *total quality management* (TQM) approach to managing a company. Whether the company is a manufacturer, a hotel operation, an entertainment facility, or a banking operation, most companies agree that the TQM approach to customer satisfaction and quality improvement involves three major components: strategic, cultural, and technical.

Figure 1.2 graphically depicts this situation. During the early stages of process improvement, it is common to think that the main sources for improvement will come from the local workforce (production workers) using control charts or other tools to monitor the quality level. These companies mistakenly think that this effort will improve quality significantly. Further along into the journey of process improvement, most companies realize that there are other tools for the local workforce to use beside a control chart.

As companies continue maturing during the process improvement journey, they learn the importance of strategic business planning by top management. Management information systems must be designed and analyzed in proper fashion, so that the company's focus is maintained to achieve the long-range objectives of the company. To gain true leverage, design the following management reports *in such a fashion that each report can be prioritized (ranked) and action items identified*. Some important reports concern scrap-warranty costs, employee surveys, customer surveys, engineering change orders, customer complaint reports, deviation requests, accuracy of process routing sheets, purchase orders, delivery schedules, productivity, and first-time quality.

This list of management reports can be measured, displayed, and reviewed by the management team. The problem is that these items are not truly actionable items. It is the challenge of management to translate these customer wants and long-range company objectives to actionable tasks. People must be assigned the responsibility and given the time to conduct this planning phase of process improvement. The company needs to begin the journey of problem *prevention* at the earliest stage in the process, instead of having the final customer *detect* the problem.

1.3 Management Levels versus Tools Used

The proper tools used and the skills needed change as we slice through the organizational structure of the company, starting with the ranks of the local workforce up through the ranks of middle management, and finally to top management. Figure 1.3 is a model showing the tools used and different skills needed at different levels in the company. This model is not cast in concrete, and there are some overlaps of skill needed and tools used.

The efforts of top management should be concentrated on:

1. Long-range planning that will ensure a competitive position in the marketplace of the future

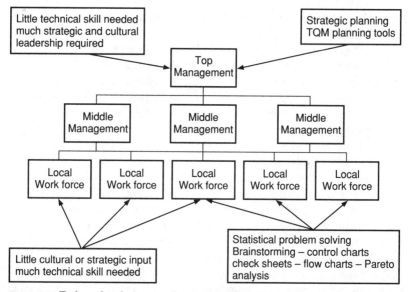

Figure 1.3 Tools and techniques to be used at different levels within the company.

2. Ensuring that the corporate culture does not conflict with the long-range objectives in the strategic plan (or at least that the conflict is minor)
3. Assigning resources to ensure that long-range objectives are achieved
4. Formally documenting the "voice of the customers" and learning their important wants
5. Ensuring that management systems are in place so that the company can manage in a *proactive* mode, not a *reactive* mode
6. Providing the focus for the company and then delegating authority and responsibility
7. Participating in training and being capable of training lower levels of management
8. Having a basic understanding of needs at all levels of the organization

Middle-level management responsibilities will require a mix of both cultural and technical aspects of the TQM process. Middle-level management efforts should focus on the following:

1. Analyze and assimilate the data in management reports, customer surveys, and employee surveys.
2. Translate the voice of the customer to measurable items, trace them back in the process, and determine whether the process is both stable and satisfactory. Often the process will not be stable or capable. Management should work toward the two goals of *stability* and *improving capability* for the important processes.
3. Optimize processes so that non-value-added activities are minimized.

Much technical skill is needed at the local workforce level. The local workforce has little say as far as the cultural and strategic elements of the long-range objectives are concerned.

1.4 Seven Stages of Quality Evolution

In the past 20 to 30 years, hundreds of corporations, many national economies, and millions of people have suffered greatly because of the emergence of *global economic competition*. Many consumer concerns are related to *quality, cost,* and *timeliness*. These competitive issues translate to the producer's *market share, productivity,* and *profitability*. For a company to be competitive in the future, management must put in place systems that on a daily basis focus on

1. Customer satisfaction
2. Continuous improvement
3. Elimination of non-value-added efforts
4. Respect for people

Most of this book will deal with the technical tools used to improve processes and a company's position regarding *quality, cost,* and *timeliness;* and their relationship to *market share* and *profitability.* Figure 1.4 shows the seven stages of quality evolution.

The *least advanced* companies are considered at stage 1. They concentrate on making sure that they ship product in specification or get an engineering waiver to approve the shipment of nonconforming

Stage 7.
TOTAL QUALITY MANAGEMENT
Strategic planning tools for management.
Methods to improve the "Quality of Management"

Stage 6.
QUALITY FUNCTION DEPLOYMENT
define the *"voice of the customer"*
in operational terms [consumer oriented]

Stage 5.
QUALITY LOSS FUNCTION
Minimize the total cost of both
producing and consuming the product [cost oriented]

Stage 4.
PARAMETER DESIGN
Product and process design optimization
Design more robust function at lower cost.
[society oriented]

Stage 3.
COMPANY-WIDE QUALITY ASSURANCE
Quality Assurance involving all
departments - design - manufacturing
- sales - finance - maintenance
[departmental systems oriented]

Stage 2.
PROACTIVE MANAGEMENT
Quality Assurance during production
including foolproofing and
Statistical Problem Solving.
[manufacturing process oriented]

Stage 1.
REACTIVE MANAGEMENT
Inspection after production, sorting
of finished products, and problem
solving activities.
[product oriented]

Figure 1.4 Seven stages of quality evolution. (*Reprinted with permission of the American Supplier Institute, Inc.*)

product. These companies spend many dollars and worker-hours inspecting, sorting, and reworking product—in other words: *fire fighting*.

Stage 2 is the level where more emphasis is put on improving the manufacturing process. It is more process-oriented. At this level, companies begin to use the basic tools of statistical problem solving and try to improve the processes (and products) by attaining *statistical control, targeting,* and *capability improvement.*

In stage 3 the company expands the use of SPS tools beyond manufacturing applications. Nonmanufacturing departments and administrative departments use Pareto analysis, brainstorming techniques, run charts, control charts, check sheets, scatter plots, and other techniques as tools to aid in the process improvement effort.

The design of the experiment, or *design optimization,* is addressed in stages 4 and 5. In stage 4 we learn how to optimize the product design or the manufacturing process. Some tools that we will use during design optimization include

Pareto analysis	Brainstorming
Orthogonal arrays	Response tables
Linear graphs	Triangular tables
Signal-to-noise ratio analysis	Tolerance design

If increased material costs or processing costs are needed to improve the product, then one way to make that financial decision is to use the technique described in stage 5. We weigh the cost of a quality improvement, if there is one, versus the benefit in making that improvement.

In stage 6, customer *wants* are identified (both the internal customer and the external customer); then the *voice of the customer* is summarized, analyzed, and prioritized. A cross-functional team is responsible for ensuring that *the voice of the customer is heard clearly and that the company takes action when needed to satisfy the customer's wants* while considering quality, cost, and time to market.

In stage 7, the planning tools of TQM are used by upper-level managers who are involved in the strategic planning activities. These tools are not limited to quality-related matters. There has been much written about the "management of quality." The tools of TQM are techniques to improve the *quality of management.*

Top management uses the planning tools of TQM to develop the long-range strategic business plan and to achieve the objectives. Numerous top management people have struggled and asked for guidance as to what kind of a statistical tool should they be using as vice president of purchasing. This is an example of top managers thinking that they should be using the technical tools of statistical problem solving. (See Fig. 1.3.)

1.5 Proactive (Prevention) versus Reactive (Detection) Style of Management

There has been much discussion about prevention versus detection management, but few people understand how to use statistical tools to aid in the transition from detection to prevention management. Many people comment that we are always going to have to "detect" things and that we cannot really do "100 percent prevention" work.

Let us draw the analogy: If we want to *prevent* heart attacks, we should monitor the important elements of the process—cholesterol levels, blood pressure, body weight, diet, lifestyle, etc.—and take corrective action when needed.

In a manufacturing situation, we want to prevent

1. Production of bad product
2. Non-value-added efforts
3. Excessive energy consumption
4. Unacceptably high downtime levels

If this is the case, then quite possibly the answer is to monitor the process by using a control chart for variables and to react to out-of-control points, which will help achieve the goal of preventing bad product.

There is a two-way communication between *detection* and *prevention:*

Detection: Wait for bad things to happen, then work on putting out fires.

Prevention: React to an early warning alarm system that will warn you *before* bad things happen.

Before we go further, we should define what is meant by "bad things." This term can mean many things to different people: customer complaints, excessive inventory, stockouts, high downtime, over-budget operations, or out-of-specification product. In most of the following discussion, we consider bad things to mean *out-of-specification product* or products or services that have *low levels of customer satisfaction.*

To have a process running in a prevention mode, we need the following:

1. The process in statistical control (stable, predictable output)
2. The process centered close to the customer's target
3. The ability to easily meet the specifications (good capability)
4. A satisfactory process

Item 4 is somewhat of a catchall of all possible problems. Usually when you have the first three items, you have a satisfactory process.

Occasionally situations will arise where the process is in control, it is centered close to the customer's target, and all the products are easily inside the specification limits, yet there are customer complaints or field failures. This would be categorized as an unsatisfactory process. This could be a case where

1. The specification limits do not relate in any way to customer satisfaction.
2. The wrong quality characteristic is being monitored.
3. The target value for the customer was not correct.

The company's focus should start out with processes in need of improvement, such as a process that is producing out-of-specification product and/or causing customer complaints. In the early stages of process improvement, we are detecting bad things (out-of-specification product and out-of-control conditions); as the process is improved and centered closer to the customer's target, there will be less and less out-of-specification product and fewer out-of-control conditions. As process improvement progresses, customer satisfaction improves.

Finally when the process has been improved so that it is in a state of near-perfect statistical control and is adjusted closer to the customer's target, there will be even less out-of-specification product or possibly none at all. In the ideal situation, if an out-of-control condition occurred, there would be no out-of-specification product produced or product that is in specification but not really very good. This is how the principle of prevention actually works.

1.6 Costs of Poor Quality

There are several things a company can do to regain the competitive edge that has been lost. An astounding amount of money and resources are wasted year after year due to the *cost of poor quality* (CPQ). This point is illustrated by an example:

> A small company had $10 million in sales over a given period. During that period, warranty costs were $1,550,000. Scrap and rework were $150,000. The appraisal (inspection) budget was $250,000, and the prevention budget was $50,000. The total cost of poor quality was $2 million (20 percent of sales). Reducing the CPQ by 50 percent would increase profit by $1 million.

An efficient way of cutting back on the CPQ is to use the tools of statistical problem solving to help identify the causes of problems and to prevent their recurrence in the future. This will reduce the CPQ, increase productivity, and improve the level of customer satisfaction.

When we look at the breakdown of the costs of poor quality, it is obvious where the opportunities for improvement are located. Reducing warranty costs should be one of the strategic objectives. The warranty costs (field failures and customer returns) might be due to a product that is both out of specification and nonfunctional, or they could be from a product that is in specification but does not function properly. Warranty costs must be sorted into logical categories so that actionable tasks can be identified and assigned. Some logical ways of sorting are by product name, product number, department at fault, machine at fault, quality characteristic at fault, customer name, or type of field application. There is essentially no prevention work being done by this company. There is a need for up-front investment to assign resources to begin the behind-the-scenes work leading to the prevention of such a great deal of waste categorized as warranty costs.

There is a possibility that a great deal of out-of-specification product is being shipped to the customer and ends up causing a field failure. This could happen if engineering waivers are granted formally, allowing out-of-specification product to go to the end customer, or, for some reason, the sorting of finished product is not taking place. There is the possibility that the actions taken to reduce warranty costs might increase the amount of internal scrap and rework. It is better to have the product identified, sorted, scrapped, or reworked in plant than to have the problem show up in the field. Remember that the objective in this example was to bring down the cost of poor quality, and it was determined that warranty costs were the big opportunity for improvement in reducing the total cost of poor quality. If the early actions of reducing warranty costs cause in-plant scrap and rework to dramatically increase, then the next phase of process improvement should be the analysis of the causes of scrap and rework so that action can be taken to improve the processes that are the main contributors to in-plant scrap and rework.

In some industrial operations it will probably be difficult to relate the CPQ in its standard format. It is not uncommon for some companies to have few true documented warranty costs. There may be a feeling that there is little scrap and rework. These are symptoms of a company not having good and accurate management information systems. Some industries are including costs that are not considered in traditional CPQ situations, such as

1. Downtime costs
2. Excessively high levels of inventory
3. A percentage of energy consumption
4. The percentage of employees performing non-value-added operations
5. Overtime worked to replace or rework products not made correctly the first time

There is no one standard format for what items should be considered in the CPQ. The main effort should be to capture major items that are truly non-value-added activities.

In another instance, a small die-casting company was having very serious quality problems that manifested themselves as high cost, low productivity, and low levels of customer satisfaction. The company had never before measured the cost of poor quality. At the end of the first quarter of gathering quality costs, these costs were summarized and reviewed by the management team. (At this stage many costs of poor quality were not being tracked.) Costs were broken down into four main categories:

1. In-plant scrap and rework
2. Customer returns
3. Inspection activities, laboratory (in-house detection activities)
4. Prevention activities

The costs associated with the first three items were discussed and reviewed. The discussion turned to prevention activities. The CPQ report showed zero dollars spent toward the prevention of poor quality. The company knew before this meeting that its method of managing quality-related issues was quite outdated. There was general agreement that the focus needed to be prevention-oriented in the future. The reported CPQ for the quarter was in the area of $400,000. An agenda of required action items was assigned to members of the management team, and the meeting was adjourned. A large poster was on display, congratulating the employees on the profit for the same quarter of the year. The profit was slightly over $250,000.

This example just reinforces the powerful negative effects of poor quality. Enlightened management understands that the big gains to society lie in improving the management of quality rather than the traditional belt-tightening and cost-cutting actions of the past.

An extremely important point should be made at this time. Use statistical problem-solving tools on the big opportunities for improvement. If we try to use these tools on small opportunities for improvement or even worse (no opportunity for improvement at all), it will cost the company money, will not improve customer satisfaction, and will create confusion, skepticism, and frustration in the workforce.

Statistical problem solving can mean many things to many different companies and people. SPS is a group of statistical tools and a *new philosophy* associated with all aspects of a company's operations, whether it be marketing, research and development, purchasing, manufacturing, engineering, production scheduling, inventory control, or maintenance.

1.7 Process Improvement Translates to Quality Improvement

In the 1950s and 1960s, there was an attempt to use these statistical tools. These efforts used *statistical quality control* techniques. The main effort of the SQC program was to monitor the quality of the product by using a control chart, expecting the quality of the product to improve. Many companies were successful at improving processes, products, and services, but there were many casualties.

The only way the quality of the product will ever improve is to improve the process that yields the product. This process improvement is the focus of statistical problem solving. The process consists of

1. Material
2. Machine
3. Method
4. People
5. Miscellaneous (environmental factors, measurement systems, etc.)

This understanding is one of the cornerstones of knowledge necessary to successfully implement the tools of SPS. Once a person in the mining industry commented that the operations are at the mercy of mother nature (earth) as a provider of the raw material. There was belief that nothing can be done about the raw material; after all, the supplier of material to the mine is the earth. In this situation the material would be considered a noise factor. Some form of design experiment is the technical tool to use. The strategy should be to work on developing a process that is insensitive (robust) to material variation.

But remember, there are other elements that make up the process. What can be done to improve the output of the process by adjusting the other elements in the process (machine, method, people, and the environmental factors)?

If we can calm down the elements that make up the process and adjust those elements closer to their targets, then and only then will the output of the process improve. Then the domino effect that the late Dr. Deming talked about begins to appear: Productivity goes up, operating costs decrease, and customer satisfaction improves.

Figure 1.5 shows the principle of process improvement. The top half of the figure represents the situation in 1989. The output of the process is generating some out-of-specification product. The nonconforming product could be scrapped, reworked, downgraded, and sold at a lower price, or an engineering waiver could have been requested and approved to use the product that is out of specification. There is a need to reduce the product variation. Four elements of the process

Introduction to Statistical Problem Solving   13

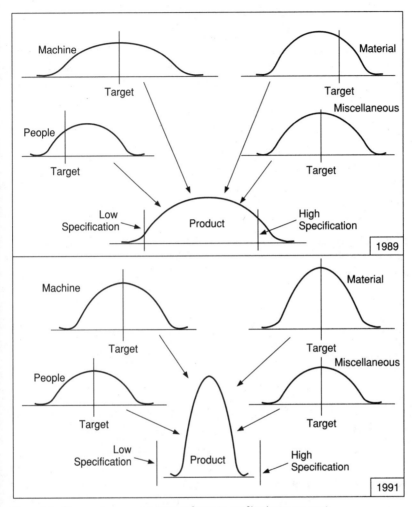

Figure 1.5 Process improvement translates to quality improvement.

are shown graphically: machine, material, people, and miscellaneous. The machine element of the process appears to be on target. The material aspect of the process is not centered on the target. The people aspect of the process is a great deal off target. Finally, the miscellaneous aspect of the process is centered. To improve the quality of the product, it is necessary to improve the elements of the process (reduce variation of the elements of the process and center the process on the targets).

The bottom portion of Fig. 1.5 represents the situation in 1991. Notice that the output of the process has improved (reduced variation and better centering on the target). This was accomplished by improving the elements in the process. The variation in the machine aspect of the process in 1991 is less than it was in 1989. The material aspect of the process has improved, variation has been reduced, and the material is centered closer to the target. The people aspect is centered closer to the target in 1991 than in 1989. The miscellaneous aspect (environmental elements of the process, measuring system, etc.) of the process appears to have changed very little.

Figure 1.5 is generic, but the principle has wide application. If we want the quality of the product to improve, the process that yields that product must be improved. These things must be done to achieve process improvement:

1. Determine which product quality characteristics translate to high levels of customer satisfaction.

2. Determine which are the important process parameters that strongly influence the process.

3. Discover what the optimum targets for the process parameters are.

4. Find a way to reduce the variation of the important elements of the process.

Quality function deployment and design of experiment are some techniques in addition to SPS that can be used to help overcome these obstacles.

1.8 Common Pitfalls Encountered during Process Improvement

The most common obstacle to successfully implementing process improvement is that the management team is not involved on a daily basis and is not committed totally. One reason why managers are not involved on a daily basis is that management is not confident of their knowledge of the skills needed to manage quality. Today senior managers are under the gun to produce short-term results, and there is the misconception that efforts to improve processes, quality, and customer satisfaction will have a negative impact upon operations. These are some of the signs of involvement by senior management:

1. Senior managers learn how to apply these techniques in their areas of responsibility.

2. Senior managers personally apply these techniques.

3. Senior managers train the workforce in process improvement.

4. Senior managers regularly attend and participate in monthly or quarterly process improvement review meetings.

If senior managers are trained, involved, and committed, then their actions will direct and support the efforts in the application of process improvement. We hear all the time that a company wants to get the best bang for its dollars. Then senior management should lead the workforce to the big opportunities for improvement. Very seldom does the improvement come overnight.

This lack of understanding, involvement, and commitment by senior management manifests itself in many ways.

1. There is not enough training or the wrong kind of training with no organized plan to use these techniques in the company's operations.
2. The quality assurance department has the great majority of the responsibility to ensure that the techniques are used successfully. The other departments are not responsible, and after all, the quality assurance personnel are the quality experts. This rationalization resulted in few of the key departments getting involved (sales, marketing, purchasing, product design, manufacturing engineering, maintenance, finance).
3. Since the quality assurance department had the prime responsibility for getting these charts and graphs out to the factory floor, the main effort became a contest to get the most charts onto the factory floor.

If the senior management team is trained properly, involved, and committed on a daily basis, this will become common knowledge throughout the workforce and this effort will be taken more seriously.

1.9 Target Value Principle

If you asked a number of people who work in industry what the word *quality* means, you would get many different responses. One of the most common definitions of quality is something like this: *Quality means consistently meeting specifications.* In other words, if the product is within the engineering specification, there is no problem. This is the definition that most of us have grown up with.

There is a more enlightened way to think about quality. The Japanese have created many interesting definitions of quality. One of the best-known definitions of quality is as follows: *Quality means making product uniformly around the target value.* This may appear to be a rather odd way to define quality. One reason that most people in the United States do not feel comfortable with this definition is that it does not address whether specifications are being met. This brings us to an interesting discussion.

What is the purpose of specifications? Many people would say, "That's how we tell good product from bad product." This type of thinking is strongly related to the *detection* style of management. Today, most people involved in design and manufacturing need to know what the specifications are because they cannot consistently meet specifications. If these people focus on the specification as a means to tell good product from bad, then much of the problem-solving activities could degenerate into arguing whether the tolerance is realistic, rather than trying to improve the process so the uniformity of the product improves around the target value.

That leads us to the next point: *What should the target value be?* Many people take the approach that unless logic in engineering information dictates differently, consider the target value the midpoint of the specifications. This is not always the best approach. Many people say, "Listen to the 'voice of the customers,'" which means, "Ask the customers where they want the process or product targeted."

> Once I was walking through a mall in Tulsa, Oklahoma. In an open area, there were three cars and a group of marketing survey people. I was asked if I would participate in this survey. I said yes. I was told to go get into the first car and perform a number of tasks: actuate the windshield wipers, power seat, warning lights, radio, and power trunk release. After completing these tasks in each of the three cars, I was asked to respond to a questionnaire relating to the perceived level of quality of the three cars. I was rewarded with a "shiny silver dollar" for my involvement.
>
> Later, I walked past the survey team and struck up a conversation. I found out that one of their major interests was to find out what door-closing effort the average customer is most satisfied with: Obviously, there is no one door-closing effort that all customers will like best. The first car was specifically assembled and adjusted with a closing effort of 8 lb, the second car with 13 lb, and the last car required a 17 lb effort to actuate the door. One of the questions asked on the questionnaire related to how the door felt when it was opened and closed. With the proper analysis of all the completed surveys, the car manufacturer can determine the target value for door-closing effort and then in turn deploy the voice of the customer in the design, processes, and finally finished products or services.

Many people will say, "The purchase contract requires only that I meet engineering specifications." In the past many contracts have been written in the above manner, but the new generation of contracts centers on a new relationship between buyer and seller. The optimum target value can be determined by a number of methods: market surveys, engineering studies, scatter plots, or design of experiments to determine the optimum target values.

Figure 1.6 represents graphically the two definitions of quality. The top half relates to the concept that *quality* means *consistently meeting engineering specifications.* Consider the shaded area as the magnitude of the cost penalty that either the producer or the consumer must pay during the life of the product. The scale on the horizontal axis represents the quality characteristic. This concept reinforces the idea that if the product is inside the specification, there is no penalty. The letters

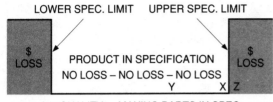

(a) "QUALITY" = MAKING PARTS IN SPEC.

(b) QUALITY = MAKING PRODUCT ON TARGET

Figure 1.6 (a) Traditional definition of quality and (b) Dr. Taguchi's definition of quality. (*Reprinted with permission of the American Supplier Institute, Inc.*)

X, Y, and Z represent three parts produced from a manufacturing process.

Products X and Y are good products, but product Z is not a good product. It is outside the specifications. It is scrap or rework or possibly requires deviation approval. But when we notice how close products X and Z are to each other, then we ask, "What is the difference between X and Z?" The answer is, "not much."

The top portion of Fig. 1.6 is a graphical example of *detection* management. The bottom portion represents the definition of quality as "making product uniformly around the target value." Again, the shaded area represents the loss to either the producer or the consumer during the life of the product. The bottom portion of the figure is most well known as the *Taguchi loss function*. As you can see, the farther a product deviates from its target value, the greater the cost penalty to either the producer or the consumer. Again, assume that X, Y, and Z represent three products produced.

How should management describe these products? Product X is the best possible product produced because it is very close to the customer's target value. Product Y is thought of as a fair product because of the distance it deviates from the target value. And finally product Z is classified as a poor product because of the greater distance it is away from the target value.

It is at this time that someone usually asks the question, "Do you want me to ship parts X, Y, and Z?" This book is not about learning to make the right decision regarding the shipment of parts X, Y, and Z. The purpose to this book is to learn techniques that will help prevent the production of parts like Z. The big opportunities are not in detection—they are in prevention! Then we will not have to make that tough decision: "Should I ship the parts?"

1.10 Summary

1. When discussing quality, one should also consider cost, timeliness, and customer satisfaction. The challenge is to balance all these factors. Management must meet the challenge head-on, starting out with a long-range strategy addressing customer satisfaction, quality, cost, and productivity.
2. There are three elements of total quality management: strategic, technical, and cultural. There is a set of management planning tools to be used to develop a logical, organized strategic plan. Management must have systems that provide feedback to verify that the improvement effort is successful. The local workforce (factory workers) needs the skills to use the technical tools of problem solving on a daily basis.
3. Work toward managing the company in a proactive rather than reactive fashion. Develop systems that will require action before trouble develops. Work toward statistical control, targeting, and capability improvement (variation reduction).
4. Use all the tools in the SPS toolbox. Some of the more common tools are Pareto analysis, brainstorming techniques, measurement system analysis, histograms, run charts, control charts, check sheets, scatter plots, and design of experiment.
5. Concentrate on reducing the cost of poor quality. If this is done properly, customer satisfaction improves, production goes up, and costs come down. Make the transition from inspection to prevention.
6. Process control not quality control. Often we will monitor the quality of the product. If we need to improve the quality of the product, we must improve the elements of the process (material, machine, method, people, miscellaneous).
7. Quality should be thought of as *making product uniformly around the customer's target.* The challenge is to identify important characteristics of the product and the process, then to find the optimum target.
8. Guard against these roadblocks to successful implementation of SPS:
    a. Management is not committed and involved.
    b. There is not enough training.
    c. Quality assurance is thought of as an inspection department program.
    d. Quality assurance tried to wallpaper the walls with control charts.

Chapter 2

Pareto Analysis and Brainstorming Techniques

The first step of the planning phase of process improvement is to identify the *vitally few* important opportunities for improvements. Once this is done, the energies of the company can be directed toward improving the processes deserving attention. A good manager uses the Pareto principle numerous times during the day. This principle must be used because of the company's limited resources.

2.1 Styles of Management

Consider the comments by two managers, Jovita and Ned:

> JOVITA: The process improvement teams must identify which product lines and processes deserve the most attention.
>
> NED: All product lines and processes must have process improvement teams assigned.
>
> JOVITA: The company must identify the important operations within the company and train those employees first in the techniques of statistical problem solving (SPS) and process improvement.
>
> NED: All the employees must be trained in the tools of SPS and process improvement and be on a team.
>
> JOVITA: The company must identify which vendors are most critical to the company's goals and work with those vendors closely to improve their products and processes, which will improve operations at our facility.
>
> NED: The company must get a printout of all our suppliers and make sure that all the vendors have process improvement teams at their facility.

The comments from Jovita indicate that she understands the need of focusing the company's efforts and resources. Jovita has informally used the Pareto principle to allocate the limited resources of the company. If Jovita does follow through on her comments, she is taking the first step toward an efficient process improvement effort.

The comments from Ned sound as if he is using the "shotgun" approach to process improvement. Ned probably assumes that the more process improvement teams the company has, the more things will improve. Ned does not seem to appreciate the need for process improvement efforts to be managed and given direction.

2.2 Pareto Analysis

There are many sources of information that can be summarized and analyzed so that the company can focus on the "big hitters" that affect cost, quality, and time to market. Some sources of information include

- Customer surveys
- Employee surveys
- Customer complaint reports
- Productivity reports
- Quality reports
- Internal cost analyses by operation
- Internal cost analyses by part name
- Profit margins by product line
- Summary of engineering concerns for new products

These are some of the more common ways of constructing a Pareto analysis:

- Scrap or rework in dollars by part number or part name
- Scrap or rework in percentage by part number or part name
- Scrap or rework by defect cause
- Number of deviation requests by part number or part name
- Dollar value of product sold at downgraded price
- Number of customer complaints by part number or part name
- Number of injuries by type or cause
- Number of incorrect shipments by cause code
- Dollar value of inventory on hand by commodity
- Amount of machine downtime by department
- Amount of machine downtime by cause code

The information which will point the company toward its vital few big problems will usually come from the financial department, possibly from the industrial engineering department, or from the scrap report that the plant generates.

History of Pareto analysis

The Pareto phenomenon was first observed by Italian economist Vilfredo Pareto. He observed that 20 percent of the population has 80 percent of the money. There are a few very rich people in the country, and there are many people considered trivially poor. Mr. J. M. Juran was one of the first to apply Pareto analysis to industrial situations. A general rule is: 80 percent of the opportunities for improvement come from only 20 percent of the possible sources of improvement.

A Pareto analysis is an extremely powerful tool that will identify the few vitally important problems. This breakdown of problems can be done manually or with the aid of a computer.

Pareto analysis example

Figure 2.1 is a Pareto analysis of scrap in dollars by department number. The analysis of this information tells us to concentrate our efforts on departments 51, 54, and 67, and not the other departments. If the company wants to make significant gains in reducing the financial drain, the efforts of the company must be directed at solving the few vital problems. Many people caught up in the heat of day-to-day fire fighting will be under the false impression that there are hundreds of

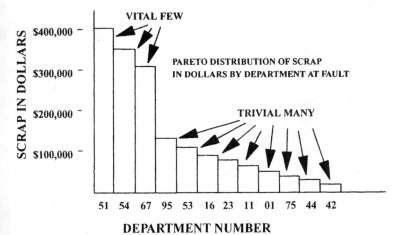

Figure 2.1 Pareto analysis of scrap in dollars by department at fault.

problems requiring attention. Do not ignore the other departments, but concentrate on the big opportunities.

Failure to prioritize problems will lead to spending short periods of time on a great many problems. People will expend a great amount of energy, become frustrated, and not see significant improvement in the situation. Instead, we should focus on the few vital problems and approach them in an aggressive, logical manner. It is not necessary for a process to be broken before process improvement efforts begin. If sorting scrap by department seems logical, then a second level of Pareto analysis must be conducted before action can be taken. Pareto analysis must also be done by part name, part number, machine at fault, and/or cause code.

> QUESTION: Is it true that some companies are required by their customers to have some form of process improvement in all aspects of their business? If this is the case, then it seems as though what you are suggesting goes against some companies' requirements.
>
> ANSWER: Responsible managers should focus their efforts where the impact will be greatest in the areas of quality, cost, productivity, and customer satisfaction. I have heard of many instances of companies requiring that their operations and those of their suppliers undergo process improvement efforts in all aspects. If there is some overriding internal procedure or a requirement in a contract that all aspects of the business be involved in process improvement, then by all means meet those requirements. The company must still focus its energies in the area that will yield the greatest opportunity for improvement and must communicate what those areas are to all employees.
>
> The identification of the few vitally important problems for the company to concentrate on is the first step in launching an effective process improvement effort. If this first step is skipped, the efforts will not result in noticeable improvements. There will be confusion and frustration. Many companies fail to conduct a thorough Pareto analysis. People become too anxious to get started using some of the more glamorous statistical tools (control charts or design of experiment).
>
> QUESTION: What do you do if the Pareto analysis does not identify the few vital opportunities for improvement? Is it possible that the sizes of all causes of problems are about the same?
>
> ANSWER: If the detective work of searching for the big opportunities for improvement is not successful, it is usually a sign of one or more of the following:
>
> 1. The data were not stratified (sorted) properly into the correct segments.
> 2. The data have been distorted so that the actual big opportunities do not appear so large. A departmental manager does not understand that the efficiency of the process improvement effort could be reduced by "fudging" the data.

3. The process could be performing as well as it can. In other words, the process is in a state of statistical control. There is only common cause variation in the process. The fact that the process is running as well as it can does not imply that the level of performance is acceptable. In this situation, a major overhaul of the process will be required. Management must make this overhaul. Workers in the trenches do not have the authority to make the kinds of changes necessary.

The first and second reasons are the most common causes of failing to identify the big opportunities for improvement. From 80 to 90 percent of the opportunities for improvement will require management action. The more involved the top management team is in the process improvement effort, the more successful the effort will be.

2.3 Data Analysis Must Yield Actionable Information

While touring a factory in southern Indiana, I noticed the "nonconforming material" area. I saw many reject tags on top of rejected material. One tag was filled out with the following information: part number, quantity, part name, and finally the reason the product had been rejected. Listed under *reason* were the following words: *parts no good—see Earl.*

If the objective of the company is to reduce the number of "no-good parts," then the management information system in this extreme case is not adequate to translate the objective to actionable tasks.

At another company's operation, the supervisor of the customer services department said that he personally investigates all customer complaints. As he was saying this, he picked up a large three-ring binder that contained all that year's customer complaints. I asked him what that stack of customer complaints told him. He responded, "We're in trouble."

The purpose of mentioning these examples is not to ridicule or belittle the parties involved. It is to emphasize we must have management information systems that can separate the vital few from the trivial many. It is very important that the data in the management information systems be designed so that the summarized analysis of the reports can help set priorities. Then the next step could be to

1. Determine cause-and-effect relationships

2. Focus on the major opportunities that will translate to significant improvement

Some management reports that should be evaluated for adequacy cover

- Scrap
- Rework
- Customer complaints
- Warranty costs
- Shipping schedule data
- Deviations
- Engineering change orders
- Job cycle time customer surveys
- Employee surveys

One approach that many companies have adopted and that is considered very *user-friendly* is the *menu system*.

A company was concerned with improving the delivery performance of its product. The company is currently tracking much information about its process on a computerized spreadsheet. There is one significant shortcoming of this report. The major reasons for late delivery of the late machines are not recorded anywhere in the report. The causes of late delivery cannot be summarized, analyzed, prioritized, or affinitized so that the management of the company can start the journey of proactive management. (Prevent late deliveries, rather than the detection of late deliveries.)

A menu of items to be input on the spreadsheet regarding the cause of late delivery must be developed and used. A possible menu might be like this:

Reason for late delivery	Cause code
Omaha late part	A
Washburn late part	B
Pittsburgh late part	C
Service center scrap	D
Service center workforce	E
Service center scheduling	F
Vendor scrap	G
Vendor late delivery	H
Trucking	I

If one inputs the applicable code, then the analysis of cause codes is made much easier and more efficient than if one had entered verbal data. An example of a spreadsheet format is shown here with the cause codes.

Job no.	Customer name	Product code	Invoice amount	Job cost	Early/late	Reason	No. days
93101	Beth	241	$50,000	$45,000	Late	A	−39
93102	Monta	241	19,500	15,000	Late	C	−51
93103	LDR	267	59,000	50,000	Early		+8
93104	TJK	238	34,500	26,000	Late	C	−19
93105	Beth	241	17,000	15,500	Early	E	+15
93106	Eloko	267	41,000	37,000	Late	C	−41
93107	LDR	708	112,500	88,400	Early	A	+10

This table represents just a short portion of the entire year's business—there were over 400 jobs during fiscal year 1993. These data must be stratified (sorted) into logical segments of the business so that the big opportunities for improvement can be accurately identified. Some logical questions to ask when planning the analysis of similar data would include these:

Should the information be sorted into different groups by product code?

Should the information be sorted into different groups by dollar value of the job?

Is there a relationship between invoice cost and number of days late or early?

Should the information be sorted by the reason for late delivery?

Pareto analysis of work order forms

A company is receiving many customer complaints. The company has found mistakes within the company resulting from the work order form not being filled out correctly.

Figure 2.2 is an example of the form that the salesperson fills out. This form also acts as a manufacturing routing sheet and is used for inventory control purposes. Each block of the order form was assigned a number. An audit was conducted of completed order forms, listing the blocks by title and frequency of each block causing a customer return.

The analysis of the work order tally sheet given by Table 2.1 tells us that block 4 (special instructions) was the part of the form that caused the most frequent mistake. Usually, this mistake was made by the sales order person recording specific customer instructions improperly or omitting the instructions entirely. The second most frequent mistake was that the steel heat number was incorrect or was omitted. The production people were not involved in the filling out of either of these blocks on the work order form.

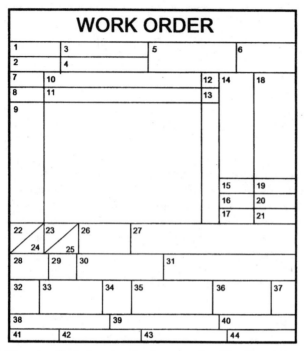

Figure 2.2 Typical work order form completed by phone salesperson.

TABLE 2.1   Summary of Work Order Mistakes by Block Number

Block no.	Frequency	Block no.	Frequency
1. Mill order no.	2	23. Invoice number	0
2. Ship to	0	24. Customer order no.	16
3. Customer no.	0	25. Authorization code/part no.	0
4. Special instructions	87	26. Date entered/by	0
5. Commodity code	0	27. Shipped date	0
6. Floor location	1	28. Driver/carrier	0
7. Date billed	0	29. Quantity ordered	21
8. Certificate code	0	30. Description	0
9. Bill of lading no./weight	0	31. Units shipped	0
10. Filled by	0	32. Quantity shipped	0
11. Operation no. 1 date	0	33. Price	0
12. Operation no. 2 date	0	34. Scrap	0
13. Operation no. 3 date	0	35. Entered by	0
14. Cost center 1	0	36. Date promised	0
15. No. 1 processing time	0	37. B-build date promised	0
16. Cost center 2	0	38. Shipped	0
17. No. 2 processing time	0	39. D-delivery date promised	0
18. Cost center 3	19	40. B/O quantity	0
19. No. 3 processing time	0	41. B/O work order	0
20. Sold to	0	42. Tag no.	0
21. Ship to customer via ___	4	43. Steel heat no.	25
22. Cut pieces	0	44. Inventory item no.	14

This example just reinforces the late Dr. Deming's position that most of the opportunities for improvement in cost, quality, and productivity will require management action. Seldom can members of the local workforce correct the situation.

Often members of the management team say that the reports they review or the quality charts their company are using do not really tell them anything. Managers say that the information must be sorted differently or that honest, detailed information is not being reported. Very little will be gained until this situation is corrected.

Pareto analysis summary

The prioritizing of the opportunities for improvement is part of the planning phase of process improvement. Upper management must be involved in this phase of the process improvement effort. Which is more important to the success of process improvement, doing the right thing or doing the thing right?

Pareto analysis helps ensure that the company does the right thing. If the Pareto analysis is not done properly, then the company is beginning a journey down a road that has little chance of success. *The Pareto analysis is an example of focused thinking.* This focused thinking is an integral component in the process improvement effort.

The prioritizing and subsequent reprioritizing of information at all phases of process improvement should be done. Comments and other information on a control chart or check sheet should be prioritized. Employee opinions as to the cause of problems should be prioritized. Reviewing the capability of numerous manufacturing processes should be prioritized. Customer survey comments should be prioritized. [Verbal information may need to be grouped into natural groupings (affinitized); then the Pareto analysis process can continue.]

2.4 Expanded Thinking

Once the Pareto analysis has been completed and the management team has reached consensus as to which things are the big opportunities for improvement, the next step is to develop an action plan to accomplish the goals. This next step requires some form of brainstorming techniques. The term *brainstorming* implies that a group of people are using their collective energies to discover something. *Remember,* when you are using these brainstorming techniques, there is not just one right answer. The success of the brainstorming phase of process improvement is dependent upon the skill, experience, and creativity of the team members. Earlier in the chapter the topic of Pareto analysis was explained. The prioritizing phase of process improvement is an example of focused thinking. The next step requires expanded thinking (brainstorming) to accomplish the following:

1. Identify the factors or elements in the process that contribute to the problem. In this situation we are detecting the problem. Brainstorming is necessary to identify a list of potential causes so that subsequent actions can be taken to improve the process, which will in turn improve the quality of the product and/or service.
2. Separate the causes of the problem from the observed effects. Use the matrix model when there is confusion or disagreement as to the causes of the problem. When a process is very complex, the matrix model technique will help simplify cause-and-effect relationships.
3. Develop a plan of actionable tasks to achieve the long-range objectives of the company. This type of brainstorming is more proactive than the brainstorming of causative factors. The how-why technique is used in a situation such as this. The how-why technique is one of the planning tools of TQM. This technique is very helpful in project management, long-range business planning, and strategic planning.
4. Develop a contingency plan to prevent unwanted, bad things from happening. Contingency planning is done during
    a. Failure-mode effects analysis
    b. Design review
    c. Quality function deployment
    d. Any activity relating to new product, process, or service development
    e. Long-range business planning

Cause-and-effect (fishbone) method

Figure 2.3 is a universal cause-and-effect diagram. We are aware of the undesirable effect in the product or service. The effect is present because of one or more causes in the process.

Remember, one of the cornerstones of understanding statistical problem solving and process improvement is that in order to improve the product or service, the process that produces the product or service must be improved. Brainstorming techniques are the primary way of improving processes. In the example shown in Fig. 2.3, the process consists of material, equipment and tooling, method, people, and miscellaneous (environment or the measurement system).

The cause-and-effect diagram is sometimes known as a *fishbone diagram* or an *Ishikawa diagram.* It is a brainstorming tool. The main purpose is to determine the causes of the problem under study. This is accomplished by generating a large list of potential causes of the effect. Do not be concerned about the validity of some of the suggested causes in the early stages of the brainstorming process.

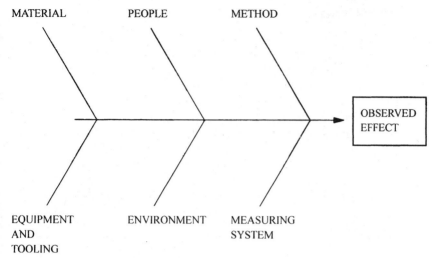

Figure 2.3 Cause-and-effect diagram (fishbone chart).

Here are some basic suggestions for use of the cause-and-effect method of brainstorming:

1. Involve people from different disciplines in generation of the list of causes.
2. Do not be critical of people's opinions.
3. Highlight the most likely causes that the team agrees upon. If there is confusion as to which factors are causes and which factors are effects, it may be necessary to use the matrix model technique. The company will not have the resources to investigate all the opinions on the list. This narrowing down of the original list is a form of prioritizing (focused thinking).
4. Keep the viewpoint positive. Focus on problem solving rather than on finger pointing.

Once a group has gone through the brainstorming phase of process improvement, the next steps are to gather information that will support or not support the list of causes generated in the cause-and-effect exercise. Depending upon the situation, there may be a need to use one or more of the following methods to gather data:

Check sheet
Run chart
Multivariable chart
Control chart
Attribute control chart

Scatter plot
Designed experiment
Customer survey
Employee survey

2.5 Matrix Model

The matrix model is a graphical technique for identifying true causes of problems so that an action plan can be developed to prevent their recurrence in the future. The matrix model should be used in the early stages of process improvement and in the later stages if needed.

Often people think they are *preventing* the *cause* of the problem, when in reality they are working on an *effect*. In Chap. 1, we stressed that the transition must be made to *prevent* the problem rather than *detect* the problem. To accomplish this goal, we must identify the true causes of the observed effect.

This graphing method efficiently helps us sort out the causes from the observed effects. This method will help us assign our resources efficiently.

The procedure begins with taking an objective or goal and then mapping out the sequential relationships between related factors. The *matrix model* begins to draw conclusions about cause-and-effect relationships.

A matrix model diagram should be used when

1. A subject is so confusing that the relationships between causes and effects are difficult for the process improvement team members to agree upon.
2. The correct plan of actionable tasks is critical to success.
3. There is a difference of opinion among the process improvement team members that the topic under investigation is really an effect and not a cause.

Example 2.1 A manufacturer of aircraft is struggling with process improvement. Team members were asked, What things do you think are hindering the process improvement effort? The following 11 items were listed:

1. Profit and schedule are more important than quality.
2. People are too busy putting out fires.
3. Process improvement cannot be cost-justified.
4. Process improvement is not understood by management.
5. Internal departments are not cooperating with each other.
6. Some people do not want it to work.
7. There is too much finger pointing.
8. Production workers will not want to fill out the charts.
9. Goals are short-term.
10. There is a lack of resources.
11. Company will not train the right people.

These 11 items will be written on a large piece of paper or posterboard at the left border, going down. The same 11 items are written along the top border, going to the right. The team members ask the question, Does item 1 from the vertical list cause item 2 from the horizontal list to happen, or does item 2 cause item 1 to happen?

Does the fact that profit and schedule are more important than quality cause people to be "too busy putting out fires"? If the answer is yes, then draw an up arrow ↑ (which is called an *out arrow*); this indicates a causal relationship. A horizontal arrow pointing to the left ← (an *in arrow*) suggests that the factor is an effect.

This exercise is repeated for the complete list of factors. This can become a very laborious task when there are a large number of factors in the matrix. If team members working on the project feel there is a need to clear their heads, then the team should take a break.

The next step is to total the in arrows and out arrows for all the factors listed along the left edge of the matrix. Then total all the in arrows and the out arrows. The totals should be identical. Once we have completed the in arrow, out arrow, and total columns, then the analysis phase can begin. If the totals are not identical, then the matrix model should be reviewed for mistakes or omissions.

Figure 2.4 is the completed matrix model. Recall that the objective of this phase of process improvement is to determine the causes of the

↑ = cause (out) ← = effect (in)	1. Profit and schedule are more important than quality	2. Too busy putting out fires	3. Cannot cost justify process improvement	4. Process improvement not really understood by management	5. Internal departments not cooperating with each other	6. Some people don't want it to work	7. Too much finger pointing	8. Production workers won't want to fill out charts and graphs	9. Short-term goals	10. Lack of resources	11. Company won't train the right people	In	Out	Total	Cause or effect
1. Profit and schedule are more important than quality		↑	↑	←	↑		↑		↑			1	5	6	C*
2. Too busy putting out fires	←			←			↑		←			3	1	4	E
3. Cannot cost justify process improvement	←			←					←	↑		3	1	4	E
4. Process improvement not really understood by management	↑	↑	↑		↑		↑		↑		↑	0	7	7	C*
5. Internal departments not cooperating with each other	←		←				↑		←			3	1	4	E
6. Some people don't want it to work												0	0	0	–
7. Too much finger pointing	←	←		←	←			↑	←			5	1	6	E
8. Production workers won't want to fill out charts and graphs				←								1	0	1	–
9. Short-term goals	←	↑	↑	←	↑		↑			↑		2	5	7	C*
10. Lack of resources			←					←				2	0	2	–
11. Company won't train the right people				←								1	0	1	–
Total												21	21	42	

What are the things that are hindering process improvement?

Figure 2.4 Matrix model chart listing some factors that hinder process improvement efforts.

situation so that an action plan can be developed to prevent recurrence in the future.

Matrix model analysis

There will usually be more than one causal factor. The team working on this assignment should take the time to verify that the analysis of the matrix model is logical.

The team has identified three factors as causes of hindering the process improvement efforts:

- Profit and schedule are more important than quality.
- Process improvement is not really understood by management.
- The goals are short-term.

After team members completed and analyzed the matrix, they noticed that the first and third causes listed could really be grouped together. Now that the team has gone through this exercise of expanded thinking, the next logical step is to focus on the causes of the factors that are hindering process improvement efforts. If top management wants to improve the process improvement efforts, then an action plan must be developed that will move aside these *cultural roadblocks* to process improvement.

> QUESTION: Isn't there a strong chance that a different team completing the matrix might come up with a different number of in and out arrows for some of the factors? It seems as though we are relying very much on the team's intuition as to what causes what.
>
> ANSWER: There is a strong likelihood that a different group would come up with different numbers of in and out arrows for some of the factors. The final conclusions of the two teams should be very similar with regard to causes. The quality and correctness of the matrix model are dependent upon a number of skills and traits of the team members:
>
> 1. Level of problem-solving skills
> 2. Level of business knowledge
> 3. Level of technical knowledge
> 4. The extent to which team members are true "team players," working to accomplish the goals for the good of the company, not just the departmental goals

The matrix model should be part of the formal presentation of the process improvement team's proposal made to senior management. If there are questions about any aspect of the conclusions, drawn from the matrix model or other charts or graphs, corrections or revisions can be made. This technique is an integral part of the management planning process.

Remember that Dr. Deming observed that somewhere between 80 and 90 percent of the improvement in quality, cost, productivity, and customer satisfaction will require the *action of management.*

This technique can be used at lower levels within the company by the workforce. These are some example applications:

What factors affect the surface finish on a ground diameter of a shaft?

What causes mistakes on job instruction sheets?

What causes excessively high levels of work in process?

What causes late delivery of repaired units to the customer?

What are the causes of customer complaints?

The Pareto principle should be applied to sort out the causes from the effects. There are no hard-and-fast rules about what percentage of the squares should be filled with arrows. Users of this tool suggest that somewhere between 30 and 40 percent is typical.

The how-why chart will help put together an action plan.

Guard against very large matrices. Matrices can become confusing, almost mind-boggling. It is suggested that no more than 15 factors be used, unless the team members are very skilled in the use of the matrix model. If the number of factors is so large that it not manageable, then try to use the "affinity" technique for grouping factors. If there is no success in reducing the number of factors by using the affinity diagram method, then team members should vote as to which factors are more important relative to the objective. The ability of the team to reach consensus is central to the success of the process improvement team.

2.6 The How-Why Diagram

The *how-why* diagram is a brainstorming technique. This diagram is sometimes called a *tree diagram* because it closely resembles a family tree chart or an organizational chart turned on its side. The finished chart helps the process improvement team make the transition from what the objective is to a means (how) of accomplishing the objective. The objective of the company is usually not actionable. The objective can be charted or measured, but direct action cannot be taken to affect the objective. The how-why diagram method assists in the brainstorming and organizing of actionable tasks that lay out an organized plan of how to achieve the objective. The action plan is shown graphically in the form of a tree, with the number of branches increasing to the right. The items listed to the right of the how-why diagram are actionable tasks.

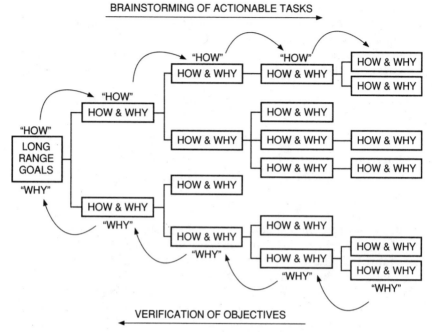

Figure 2.5 The how-why diagram. [*Reprinted with permission from S. Mizuno (ed.), Management for Quality Improvement. English translation copyright ©1988 by Productivity Press, P.O. Box 13390, Portland Oreg. 97213-0390, (800) 394-6868.*]

Constructing the how-why diagram

1. Write the goal or objective on poster paper or something similar so that it is easily visible to team members.
2. Begin to brainstorm a list of action items to achieve the objective. Write one action item (how) on a 5 × 7 in card or a sheet of paper: what can be done to accomplish the objective or how will the company accomplish it.

Usually a second, third, and sometimes fourth level of *how* items, secondary means, and possibly some tertiary levels of means are needed to ensure that the goal or objective is achieved. In this situation, the primary means become the goal of the secondary means, as shown in Fig. 2.5.

Example problem

A long-range goal of the company is to reduce rework by 50 percent within the next 2 years. The company has been measuring rework for the last year (see Fig. 2.6). Little improvement has been observed. A process improvement team decided a more formal plan to accomplish the goal was needed.

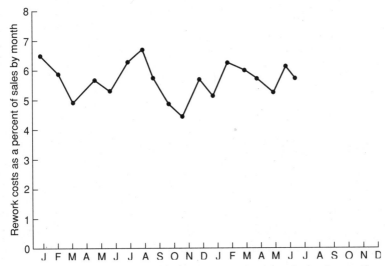

Figure 2.6 Run chart showing rework cost as a percentage of sales, by month.

How is the company going to reduce rework by 50 percent in the next 2 years? The suggestions were written on individual pieces of paper so they could be moved around. The following four items were considered *primary* ways of accomplishing the goal:

1. Provide incentives for operator to improve quality.
2. Make sure that the equipment and machinery are maintained properly.
3. Isolate the problem at the cost center.
4. Improve processes.

Figure 2.7 shows the goal of reducing rework and the four first-level methods (how's) of accomplishing this goal. These four how's are still not really actionable. The next step is to focus on the four first-level how's and ask how we will accomplish these four things. The team brainstormed a list of second-level how's. The second-level list of how's is as follows:

1. Provide incentives for operator to improve quality.   (Primary)
    A. Develop and initiate recognition/reward program.   (Secondary)
2. Make sure that the equipment and machinery are maintained properly.
    A. Rework the tooling.
    B. Develop a system to identify and correct the cause of the problem.
3. Isolate the problem at the cost center.

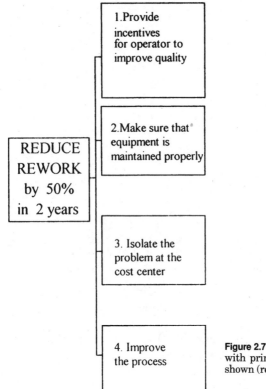

Figure 2.7 The how-why diagram with primary (first-level) means shown (reduction of rework).

    A. Detect the rework earlier.
4. Improve processes.
    A. Implement statistical problem solving throughout the company.
    B. Provide better work instructions.
    C. Do capability evaluation to prioritize problems.

At this stage, the team has brainstormed seven second-level ways (how's) of accomplishing the goal of reducing scrap. Figure 2.8 shows the how-why chart at this stage of development. The process improvement team considered item 1A as an actionable task. The other six second-level how's are not yet actionable. The team must now consider the six how second-level items and ask, How will the company accomplish these six items? The following is the list of third-level ways (how's) of accomplishing the goal of reducing scrap.

1. Provide incentives for operator to improve quality.
    A. Develop and initiate recognition/reward program.
2. Make sure that the equipment and machinery are maintained properly. (Primary)

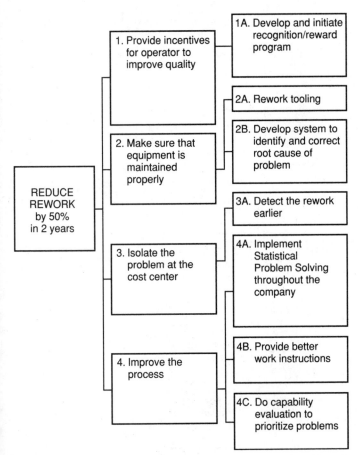

Figure 2.8 The how-why diagram with primary and secondary (second-level) means shown (reduction of rework).

    A. Rework the tooling. (Secondary)
       (1) Budget for tool repair. (Tertiary)
    B. Develop a system to identify and correct the cause of the problem.
       (1) Improve feedback report to the department responsible.
       (2) Improve the preplanning process.
3. Isolate the problem at the cost center.
    A. Detect the rework earlier.
       (1) Supervisor will be more responsible for rework reduction.
4. Improve processes.
    A. Implement statistical problem solving throughout the company.
       (1) Train engineers, supervisors, operators about the tools of SPS.
    B. Provide better work instructions.
       (1) Assign process engineer to upgrade work instructions.

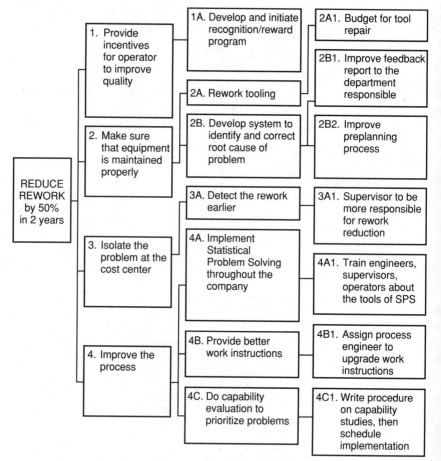

Figure 2.9 The how-why diagram with primary, secondary, and tertiary (third-level) means shown (reduction of rework).

C. Do capability evaluation to prioritize problems.
   (1) Write procedure on capability studies, then schedule implementation.

The process improvement team considered items 2A(1), 4A(1), 4B(1), and 4C(1) as actionable tasks and chose not to pursue these "branches of the tree" in greater detail.

There is still a need to continue the brainstorming process for items 2B(1), 2B(2), and 3A(1).

Figure 2.9 shows the how-why diagram with the third level of how's. The team continued the brainstorming process, asking, How is the company going to accomplish the third level of how's? Figure 2.10 is a list of four-level how statements.

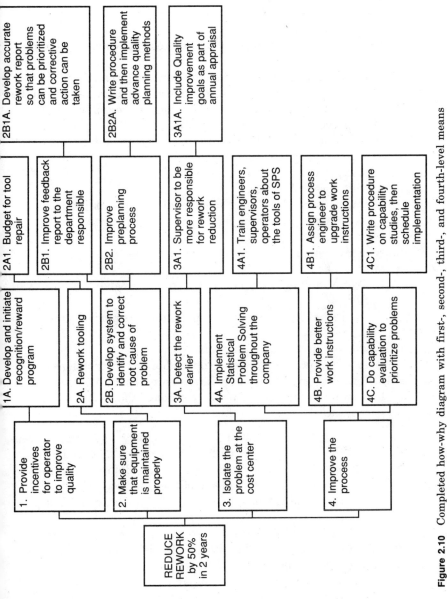

Figure 2.10 Completed how-why diagram with first-, second-, third-, and fourth-level means shown (reduction of rework).

1. Provide incentives for operator to improve quality.
   A. Develop and initiate recognition/reward program.
2. Make sure that the equipment and machinery are maintained properly.
   A. Rework the tooling.
      (1) Budget for tool repair.
   B. Develop a system to identify and correct the cause of the problem.
      (1) Improve feedback report to the department responsible.
         (A) Develop accurate rework report so that problems can be prioritized and corrective action can be taken.
      (2) Improve the preplanning process.
         (A) Write procedure and then implement advance quality planning methods.
3. Isolate the problem at the cost center.
   A. Detect the rework earlier.
      (1) Supervisor will be more responsible for rework reduction.
         (A) Include quality improvement goals as part of annual appraisal.
4. Improve processes.
   A. Implement SPS throughout the company.
      (1) Train engineers, supervisors, operators about the tools of SPS.
         (A) Include quality improvement goals as part of annual appraisal.
   B. Provide better work instructions.
      (1) Assign process engineer to upgrade work instructions.
   C. Do capability evaluation to prioritize problems.
      (1) Write procedure on capability studies, then schedule implementation.

The process improvement team felt that the how-why diagram was complete and that there were a sufficient number of actionable tasks identified to achieve the goal of reducing rework. Before the plan is finalized, the team should test the logic of the completed chart. This is done by starting at the extreme right of one of the branches and asking the question, Why?

Consider item 1A. Why should the company develop and initiate a recognition/reward program? The item to the left should answer that question. The answer is: Provide incentives for operator to improve quality. Now ask, Why should the company provide incentives for operators to improve quality? The answer should be the item to the left, which in this case is the original goal of reducing rework by 50 percent within 2 years. This method of checking the logic of the diagram should be applied to all the items. If the team feels that there is something missing or that items are not related, then the diagram should be revised. Figure 2.10 is the completed how-why diagram that the team agreed to.

> QUESTION: Why is it necessary to graph this complicated example? Most managers go through this exercise informally when a project is this complex. The only difference is that it is not documented.
>
> ANSWER: I believe that your belief that most managers go through this exercise is wrong. The planning phase of process improvement is the most important phase of all. The planning phase is the phase that needs the most attention. For discussion purposes, let's say that the managers have done this informally in their heads. Then I suggest that they take just a little time to document their thoughts formally and have other individuals review the plan.

The how-why chart is an excellent planning tool that almost forces people to really look in the mirror and ask, How is the company going to accomplish its long-range goals? This technique can be used in many situations relating to long-range and intermediate-range goals, such as

Increase market share by 40 percent in the next 2 years.

Reduce total cost of the product by 30 percent in the next 3 years.

Improve the level of customer satisfaction by 25 percent in the next 2 years.

The how-why diagram is not intended to get a company out of trouble overnight.

2.7 Gantt Chart

Once the how-why diagram is complete, the next step is to assign to individuals the actionable tasks, with planned start and finish dates. Figure 2.11 is an example of a Gantt chart with tasks and their planned start and finish dates. The Gantt chart is part of the documentation that would be kept by the process improvement team. The Gantt chart is not a brainstorming technique; it is more a method of tracking or bookkeeping relating to project management. The Gantt chart is shown because usually it is the next step following the how-why diagram in process improvement.

The how-why diagram visually displays the action items necessary to achieve specific goals and objectives. This technique helps clear up confusion about which tasks must be assigned so that the goal is achieved. Many people are caught up in the daily fire-fighting activities and lose sight of the primary means and objectives. Using the how-why diagram minimizes this difficulty.

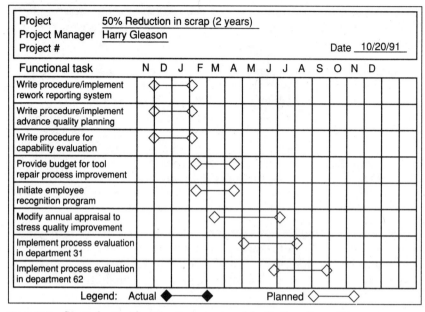

Figure 2.11 Gantt chart with action items assigned (reduction of rework).

2.8 Contingency Planning

Process improvement efforts usually require significant changes to current processes or possibly institution of a completely new and different process. Contingency planning is nothing more than documenting changes to a process, listing the concerns or vulnerabilities, and finally planning action by the process improvement team to prevent these concerns from materializing. Some people involved in process improvement seem to have a mental block when it comes to suggesting radical changes to a current process. Often the individual is spending mental energy on these concerns. This will stifle creative thinking. Everyone involved in process improvement must have training in contingency planning and other brainstorming techniques. There is always the chance that there could be some adverse effect due to process changes. One way to minimize the likelihood of these concerns materializing is to address them formally and plan them out of the new process.

It is common to have more than one concern for every process change and more than one corrective action plan for each concern. It is much less costly to consider what things could go wrong and to develop an action plan on paper. Evaluate the possible concerns and take corrective action as warranted, as opposed to making the process changes and then detecting the adverse effects of those process changes.

CONTINGENCY PLANNING WORKSHEET		
PROCESS CHANGE	CONCERN / VULNERABILITY	ACTION PLAN

Figure 2.12 Contingency plan work sheet.

Figure 2.12 is an example of a contingency planning work sheet. To complete the contingency planning work sheet, do the following:

1. List one of the proposed changes to the current process in the first column.
2. List the concerns or vulnerabilities that concern the team if these changes were made. There will probably be more than one concern or vulnerability. Each concern should be listed and numbered separately in the second column.
3. List the team's action plans to prevent these concerns or vulnerabilities from being realized. Usually there is more than one action plan item for each of the concerns.

2.9 Summary

For process improvement efforts to be successful, a combination of focused thinking (prioritizing) and expanded thinking (how-why) is essential. The management team must determine what are the major opportunities for improvement in cost, customer satisfaction, productivity, product development time, and quality. Are there specific segments of the market needing the most attention?

The tools described in this chapter are not mathematical tools but rather planning tools. Management summary reports must be detailed enough so that the company's efforts achieve the objectives. Once the specific areas needing attention have been identified by using techniques to focus the company's efforts, some form of brainstorming techniques (expanded thinking) must happen. The ideas developed during the expanded thinking phase will be the action items necessary for process improvement.

The management team must guard against the "shotgun" approach to process improvement. The focus of the company will be blurred unless there is a prioritization of opportunities for improvement.

- Strive to find the deep systemic processes needing improvement.

Simultaneously work toward:

- Strongly considering the "up-front" processes. These processes are influenced by top management, sales, marketing, purchasing, and product development. These processes will have a multiplying effect because they affect downstream processes which increase costs and delays.

- Track the elements of the process back to the root cause of the observed effect under study.

- Consider using the matrix model technique when there is disagreement if an item is a cause or an effect.

- Use the how-why diagram technique to develop an action plan for the objective under study. Once the diagram is believed to be correct and complete, then check the logic of the completed diagram by working backward (starting at the extreme right of the diagram). Ask the question: "Why do I want to take this action?" The answer should be the phrase to the left. Repeat this exercise until you have moved to the original objective.

- Use the contingency planning work sheet to initiate necessary action plans which will prevent potential problems from happening with the new process.

Chapter 3
Introduction to Variation and Statistics

The objective of this chapter is to provide the reader with (1) a hands-on working knowledge of how to interpret information expressed statistically; (2) knowledge of how to take raw, unorganized data and analyze them in proper statistical fashion; and (3) knowledge to determine whether action on the process is needed. This chapter will lay a foundation so that the reader can easily understand Chap. 5 (Shewhart control charts for variables) and Chap. 6 (capability). The number crunching that is explained in the early sections of this chapter is becoming less and less important, given the abundance of hand-held calculators and personal computers. The goal is for the reader to be able to think statistically and understand the situation in both analytical and graphical ways.

3.1 Variable Measurements

Whenever the output of a process gives continuous numerical values, that type of measurement is classified as a *variable measurement*. Examples of variable-type measurements include 0.0025 in runout, 245°C, an 84 percent machine efficiency, 15,400 lb/in^2 tensile strength, 3.22 min, 2560 ft^3/min, 37 ft · lb of torque, 4.063 in, 3.7 percent soda ash. These types of measurements are classified as *continuous variable measurements*. The prevention principle is easier to apply when the quality characteristic is a variable measurement rather than an attribute measurement.

Example 3.1 The temperature of a petroleum-based product is being monitored and recorded. Due to the limitations of the measuring instrument, temperature readings are recorded to the nearest degree Fahrenheit.

If the instrument displays 208°F and then 20 min later 209°F, probably the readings were not *exactly* 208 and 209°F. With a more sensitive measuring instrument displaying temperatures to two decimal places, we could have determined

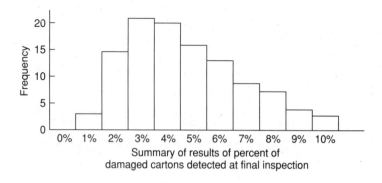

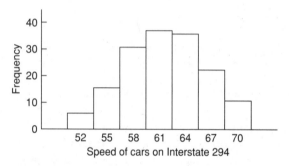

Figure 3.1 Distribution examples (percentage of cartons damaged at final inspection, speed of cars on Interstate 294).

the temperatures to be 207.82 and 209.14°F. Measuring instruments should be graduated so that there is a minimum of 10 increments within the tolerance.

When a manufacturing process or some type of repetitive activity produces numerous readings and the readings are summarized, some kind of pattern of variation will emerge. These patterns of variation are called *distributions*. Typical examples of common distributions are shown in Fig. 3.1.

3.2 Methods of Evaluating Variable Data

There are many ways of evaluating raw variable data. Some of the more common methods are as follows:

1. Calculate the average, median, range, and/or standard deviation for the data.
2. Construct a histogram to see a graphical picture of the distribution.
3. Plot the data on a simple run chart, so the data can be seen in chronological order; then evaluate the run chart visually for abnormalities.

4. Plot the data on some type of Shewhart control chart (individual and moving range chart, or possibly an average and range chart). Calculate the control limits to determine the extent to which *special-cause variation* is influencing the process, so that the special causes of the variation can be prevented from happening as frequently in the future. Then a strategy for reducing the *common-cause variation* can be developed, if needed.

None of these methods are *wrong*. It is just that some are more powerful than others, and some lend themselves more than others to certain types of processes.

Math calculation

Method 1 is the most common method of data analysis conducted. Over the long run this method will not help very much at all in our quest for process improvement. The main weakness of method 1 is twofold:

1. There is *usually* an assumption that the shape of the distribution is a normal distribution. Many times this is an incorrect assumption.
2. This method will not sort out common-cause variation from special-cause variation. One of the foundations of process control is the ability to sort out these two types of variation so that the proper action plan can be developed. *Note:* Chapter 5 will discuss common- and special-cause variation in great depth.

Histogram

An advantage of method 2 (histogram) over method 1 (math calculations) is that the shape of the distribution of individual readings can be seen. This will allow us to see if the distribution is abnormal, unnatural, or undesired. A histogram is a good method to use if the company is just beginning to introduce statistical problem solving.

Run chart

A *run chart* displays measurements that are recorded in production sequence. This charting method will show trends in the process, process shifts, or other unnatural patterns of variation. A run chart is not as powerful as Shewhart control charts for variables.

Shewhart control charts for variables

This method of data analysis contributes the most to process improvement over the long run. A word of caution should be noted here: In some situations, control charts are not really the most efficient method of data analysis. There are two situations in which control charts are not appropriate:

1. A company is in the beginning stages of gathering data from many departments, machines, and assembly lines; or numerous quality characteristics are measured from the same part. Some sort of check sheet along with Pareto analysis of the data would be the proper technique. The Pareto analysis and other brainstorming techniques will provide guidance as to where in the process a control chart or other problem-solving technique should be used.
2. The objective is to reduce the variation *within* the product, rather than reduce the variation from product to product, e.g., reducing the thickness variation along the width of corrugated board or reducing the eccentricity of a ground diameter of a shaft.

All these techniques of data analysis (math calculations, histograms, run charts, and control charts) will be shown on a case study basis. But first we do the calculations for the statistics of the normal distribution. If the reader is a novice, be patient to see the application and interpretation of these statistics.

3.3 Measures of Central Tendency

The first family of statistics, sometimes called *measures of central tendency,* tells us where the center of the distribution is located. This family of statistics tells us where the process is *targeted.* The two most common statistics used to pinpoint the center of the process are the *average* (sometimes referred to as the *mean*), denoted by $\overline{X}$ (pronounced "X bar"), and the *median,* denoted by $\widetilde{X}$ (called "tilde").

Average

The formula for the *average* is

$$\overline{X} = \frac{\Sigma X}{n}$$

where $\overline{X}$ = average
Σ = sum of readings
X = individual reading
n = sample size (number of readings)

Median

The *median* is the reading in the middle of all the readings in the sample. *Caution:* The readings must be listed in ascending or descending order.

Both the average and the median do a good job of telling us where the center of the process is located. Then we can compare the average or the median with the customer's target value so that a decision can be made whether the process should be adjusted (retargeted). If the production workers from the factory floor will be recording and plotting the readings on some type of process control chart, it *might* be easier to calculate and plot the median instead of the average. When you use the median, the sample size should be an *odd* number. It is easier to find the number in the middle if there are an odd number of readings (three or five).

Example 3.2 Octane readings from a petroleum product produced from the same refinery unit were measured from seven different storage tanks: 82.40, 82.15, 82.50, 82.65, 82.55, 82.45, and 82.35. The average is

$$\overline{X} = \frac{\Sigma X}{n} = \frac{577.05}{7} = 82.44$$

To determine the median, arrange the readings in ascending order: 82.15, 82.35, 82.40, 82.45, 82.50, 82.55, and 82.65. The median is 82.45. In this example, both the average (82.44) and the median (82.45) are quite close to each other. This is usually the case if the measurements come from a distribution that has the general shape of the normal curve and if the sample size is large enough.

3.4 Measures of Dispersion

The second family of statistics is *measures of dispersion*. The two most common measures of dispersion are the *range R* and *standard deviation* σ. The range and the standard deviation tell us the amount of variation in the output of the process. The amount of variation indicates the capability of the process to meet specifications (Chap. 6 explains the topic of capability in great detail).

Range

The range is a very simple statistic to calculate. Subtract the lowest value of the measurements from the highest value of the measurements. The symbol for the range is R. The formula is

$$\text{Range } R = \text{highest} - \text{lowest}$$

The range for the octane measurements is

$$R = 82.65 - 82.15 = 0.50$$

Standard deviation

The standard deviation describes the dispersion (variation) among the measurements. The common term for standard deviation is *sigma,* and the symbol is σ. Shown below are two of many longhand formulas used to calculate the standard deviation:

$$\sigma_{n-1} = \sqrt{} \quad \text{or} \quad \sigma_n = \sqrt{}$$

If there are fewer than 25 measurements, use the formula on the left. Use the formula on the right if there are more than 25 measurements. The formula on the left uses the $n - 1$ weighting as a correction factor for small sample size.

Today very few people will need to calculate the standard deviation by hand. Most everyone will have a personal computer or at least a calculator to do the necessary calculations. The example below shows the mechanics of calculating the standard deviation.

Example 3.3 Octane measurements were taken from seven storage tanks containing gasoline produced from the same refinery unit. The specification is 82.0 minimum. The readings from the laboratory were 82.40, 82.15, 82.50, 82.65, 82.55, 82.45, and 82.35.

X	$X - \overline{X}$	$(X - \overline{X})^2$
82.40	$82.40 - 82.44 = -0.04$	$0.04^2 = 0.0016$
82.15	$82.15 - 82.44 = -0.29$	$0.29^2 = 0.0841$
82.50	$82.50 - 82.44 = 0.06$	$0.06^2 = 0.0036$
82.65	$82.65 - 82.44 = 0.21$	$0.21^2 = 0.0441$
82.55	$82.55 - 82.44 = 0.11$	$0.11^2 = 0.0121$
82.45	$82.45 - 82.44 = 0.01$	$0.01^2 = 0.0001$
82.35	$82.35 - 82.44 = -0.09$	$0.09^2 = 0.0081$

$\Sigma X = 577.05$ $\Sigma(X - \overline{X})^2 = 0.1537$

$$\overline{X} = \frac{\Sigma X}{n} = \frac{577.05}{7} = 82.44$$

$$\sigma_{n-1} = \sqrt{\frac{\Sigma(X - \overline{X})^2}{n - 1}} = \sqrt{\frac{0.1537}{6}} = 0.1599$$

For simplicity's sake, the standard deviation σ is rounded to 0.16. Figure 3.2 shows graphically how we should interpret the octane readings with an average of 82.44 and a standard deviation of 0.16. The area under the normal curve is 100 percent. Until the last 10 years or so, many people considered the normal distribution to be 6 standard deviations wide ($\overline{x} \pm 3\sigma$). The theoretical percentages of the area under the normal curve in relation to standard deviation are shown. In theory, the curve continues out infinitely.

Note: The standard deviation very accurately describes the shape of the distribution *if* the distribution is the general shape of the normal distribution. If the readings come from a distribution that has a shape greatly different from that of the normal distribution, then the standard deviation is not as meaningful and could possibly be misleading. Some more important points to remember about the standard deviation are the following:

Introduction to Variation and Statistics 51

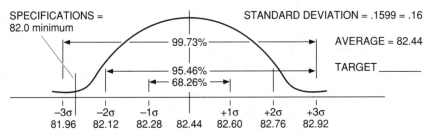

Figure 3.2 Octane readings (theoretical area under the normal curve).

1. Where is the standard deviation on the normal distribution curve?
2. What does it tell us?
3. How is the standard deviation calculated?

Remember, the average or the median tells where the *center* of the distribution is located whereas the range and the standard deviation measure the *variation* within the distribution. *The goal is to have our processes on target and very little variation, meaning a small standard deviation.*

With an average of 82.44, and a standard deviation of 0.16, we should expect 68.26 percent of the measurements to be between 82.28 and 82.60. And 95.46 percent of the measurements will be between 82.12 and 82.76. We can expect that 99.73 percent of the measurements will be between 81.96 and 82.92.

Notice that the extreme left portion of the tail of the curve is below the specification of 82.0 minimum. Many times it will be necessary to determine the portion of product outside of specification in order to make a financial decision. (This calculation, like the other calculations in this chapter, is *detection*-oriented; we are detecting the proportion of the product that is out of specification.) One goal of process improvement is to ensure that processes are running in a *prevention* mode. This is a never-ending challenge; but keep in sight the goal of detecting less and less bad product, which conversely means the prevention of bad product.

QUESTION: How should a person react if the percentage of actual observed measurements differs greatly from the predicted theoretical value?

ANSWER: One time a supervisor of receiving inspection called me at home in the evening. He said that the inspector measured the required number of bearings, calculated the average and standard deviation, and then sketched out what the normal distribution appeared to be, based upon the calculations. Of the 32 bearings measured it was expected that 22 measurements would be within $\overline{X} \pm 1\sigma = 32 \times 0.68 = 21.76$ bearings. The inspector reviewed the observed measurements—there were 24 bearings in the $\overline{X} \pm 1\sigma$ interval—and thought something was wrong. The 24 bearings in the interval would be 75 percent, not the theoretical 68 percent.

It is important to be able to use common sense when you are analyzing statistics. There is not much difference between the theoretical and the actual. The difference from the theoretically expected value could be due just to the luck of the draw in the sample, or perhaps the distribution of the entire lot is not exactly the shape of the normal distribution. There were no other conflicts in the statistical analysis concerning the bearings. There will be very few cases where the actual and the predicted theoretical values will match perfectly.

3.5 Z Score

The next application of variation and statistics requires an understanding of the Z formula and how to use the Z table. The Z formula is

$$Z = \frac{X_i - \overline{X}}{\sigma} = \frac{82.0 - 82.44}{0.40} = -2.75$$

where Z = number of standard deviation units from center of distribution to point of interest
X_i = value of interest
$\overline{X}$ = average
σ = standard deviation

Using the octane readings from the previous example, we now add the specification of 82.0 minimum to Fig. 3.3. We see that the specification limit is back (to the left) 2.75 standard deviations from the center of the distribution of octane readings.

Recall that the average $\overline{X}$ = 82.44 and standard deviation σ = 0.16. The minimum specification was 82.0.

To find the proportion of product that is below the specification of 82.0 minimum, turn to the Z table (at the end of this chapter). Go to the left edge and find 2.7, then go to the top edge and find x.x5; and where these two coordinates intersect, you will find 0.0030, which is 0.3 percent. This is shown graphically in Fig. 3.3.

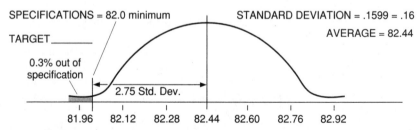

Figure 3.3 Octane readings (specification located at -2.75σ, 0.3 percent out of specification).

The great majority (99.73 percent) of the area of the normal distribution is between $\overline{X} + 3\sigma$ and $\overline{X} - 3\sigma$. Actually the normal distribution in theory continues out an infinite number of standard deviations from the center of the distribution. From a practical point of view, the important points to be aware of are the following:

$$\overline{X} \pm 1\sigma = 68.26\%$$

$$\overline{X} \pm 2\sigma = 95.46\%$$

$$\overline{X} \pm 3\sigma = 99.73\%$$

$$\overline{X} \pm 4\sigma = 99.994\%$$

The portion of the distribution that is between -3σ and $+3\sigma$, in theory, has 99.73 percent of all the items. Notice that the fourth line above shows the portion of the distribution that is between $+4\sigma$ and -4σ from the center of the process.

The importance of this point is shown in Fig. 3.4. Three factories are making the same product, and all three factories have measured the same quality characteristic from a number of products. The average and the standard deviation from all three factories were calculated. Figure 3.4 represents the output of those three processes based upon the calculated average and standard deviation. Suppose that you buy one product from each of the three factories, and those products just happen to be a part that is at the extreme tail of the distribution.

Which of the three factories will have the highest level of customer satisfaction and probably the lowest cost? Factory A will have the lowest level of customer satisfaction and probably the highest cost, since there is nonconforming product that is produced. Graphically, the specifications are $\pm 2\sigma$ from the center of the process. Recall that 95.46 percent of the product is in specification. About 4.5 percent of the product is nonconforming. These products might be scrapped or reworked, or possibly an engineering waiver would be granted to allow these products to be shipped to the customer.

Factory B will have a higher level of customer satisfaction than factory A; also since less nonconforming product is produced, we think that costs would be lower. Graphically it appears that 99.73 percent of the product is in specification, so about 0.25 percent of the product is nonconforming.

Factory C is meeting the requirements of the specification with some room to spare. Graphically it appears that the specifications are $\pm 4\sigma$ from the center of the distribution. Therefore theoretically 99.994 percent of the product is in specification, which means that 0.006 percent is out of specification. Factory C will have the highest level of customer satisfaction.

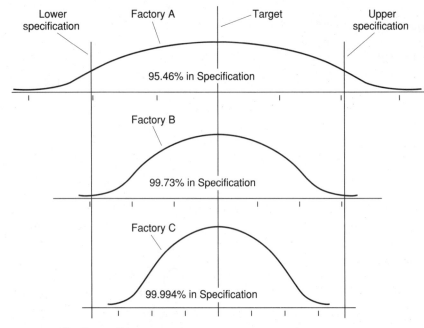

Figure 3.4 Continuous improvement.

In summary, the competitive gains in going from the level of quality at factory A to that in factory B are

1. Reduced portion of nonconforming product produced
2. Higher levels of customer satisfaction

What are the gains in going from the quality level in factory B to that of factory C?

1. Even higher levels of customer satisfaction than in factory B
2. Very minor savings in the reductions of nonconforming product produced
3. Possibly the chance to change manufacturing or assembly operations to take advantage of cost reductions or to increase productivity

QUESTION: Suppose someone from factory A does a 100 percent sort of the product, and only the product that is in specification is shipped to the final customer. In that case, shouldn't the levels of customer satisfaction be the same from all the factories? And is it not possible that factory C is spending too much money to achieve consistency and will not be competitive with factory A or B concerning price?

ANSWER: Sorting of 100 percent of the product to guarantee conformance to the specification adds cost to the product and very seldom is 100 percent effective. Sorting does nothing to improve a process. Suppose that the tails of the distribution from factory A that are out of specification are scrapped or reworked. There will still be a much higher number of products just barely inside the specifications. This will result in some amount of lower customer satisfaction.

As to the second question, possibly factory C has spent too much money achieving consistency. I am sure that there are a few situations where this could be seen as spending a dollar to save a dime. More and more contracts between producer and consumer stipulate quality levels at least as good as, if not better than, those described by factory C. Truly advanced companies consider cost as they work toward process improvement. These advanced companies understand that process improvement, done properly, leads to cost reductions, productivity improvements, and higher levels of customer satisfaction.

3.6 Histograms

A histogram is a basic statistical tool that graphically shows the shape of the distribution of variable measurements. A histogram is a good tool to use if a company is just beginning to introduce the use of statistical tools.

In this section, we consider the following:

- The best way to visually present summarized raw measurements so that the histogram gives the viewer a proper perspective of the information
- Number of cells for the histogram
- Cell width for the histogram
- Determination of cell midpoints, lower boundaries, and upper boundaries
- Construction and analysis of histogram

Figure 3.5 shows the elements of a histogram. Some conventional rules apply to constructing a histogram:

> The cell width must remain constant in all the cells of the histogram.

> The values for the upper and lower cell boundaries should be carried out one decimal place further than the measurements are carried out. Doing this will minimize any confusion about which cell a measurement belongs in.

Case study

An ordnance manufacturer is in the early stages of using statistical analysis to evaluate product performance. One performance charac-

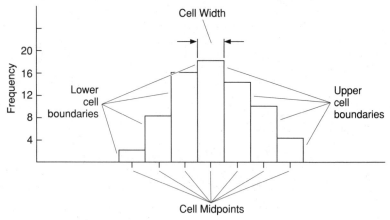

Figure 3.5 Elements of histogram.

teristic is projectile velocity. The engineering specifications are 1100 to 1200 ft/s. The target velocity is 1150 ft/s.

One sample of munitions was tested from one lot, and the velocity of 40 rounds of ammunition was recorded. Those readings are as follows:

1143	1171	1157	1138	1177	1165	1152	1133
1161	1147	1158	1136	1117	1149	1169	1151
1128	1186	1160	1152	1143	1149	1172	1158
1151	1134	1157	1148	1163	1182	1158	1171
1138	1123	1167	1159	1144	1150	1179	1129

In constructing a histogram, the histogram must have the proper number of cells. Here are some *general guidelines* to follow:

Number of cells in histogram. The square root of the number of measurements is a good approximation of the number of cells needed in the histogram. Therefore,

$$\text{Number of cells} = \sqrt{n} = \sqrt{40} = 6.3$$

The number of cells should be close to 6. Most likely we will end up with 5, 6, 7, or 8 cells in the histogram, depending on other considerations.

The next step in constructing a histogram is to determine the range of the measurements.

$$\text{Range} = 1186 - 1117 = 69$$

Different options should be considered now. The range of readings should easily fit between the lower boundary of the first cell and the upper boundary of the last cell. Five options of the trial-and-error method are shown.

Option	No. of cells	Cell width	Range covered
A	5	14	70
B	6	12	72
C	7	10	70
D	8	9	72
E	8	10	80

The best situation of the five options is E.* In options A, B, C, and D, the range covered is too close to the observed range. It might be difficult to select lower boundaries, upper boundaries, and cell midpoints for this set of measurements and yet keep the histogram easy to understand.

The cell midpoints are usually displayed while the lower and upper boundaries are not displayed in the histogram. Again, trial and error yielded the cell boundaries and midpoints shown here.

Lower boundary	Upper boundary	Cell midpoint	Frequency
1115.0	1124.9	1120	2
1125.0	1134.9	1130	4
1135.0	1144.9	1140	6
1145.0	1154.9	1150	9
1155.0	1164.9	1160	9
1165.0	1174.9	1170	6
1175.0	1184.9	1180	3
1185.0	1194.9	1190	1
			$n = 40$

The completed histogram (Fig. 3.6) visually shows that the process is well centered on the target of 1150 ft/s. We make this statement just by visually analyzing the histogram. As we look at the variation between the lower boundary of the first cell of the histogram and the upper boundary of the last cell, we see that almost all the specification width (100 ft/s) is used up. This tells us that the process is not *easily* meeting specifications—the process is *barely* meeting specifications. Recall that to have a process running in a prevention mode, one of the requirements is the ability to *easily* make product within specifications. This process would be considered marginal at best. An action plan to reduce variation of the velocity of the projectile should be developed and implemented.

The objective of Sec. 3.6 is to familiarize the reader with the steps needed to construct, complete, and analyze a histogram.

---

*These decisions come with experience and are prerequisites to understanding the fundamentals of variation and statistics, which are prerequisites to understanding process control techniques, which is a prerequisite of applying process improvement techniques in industry.

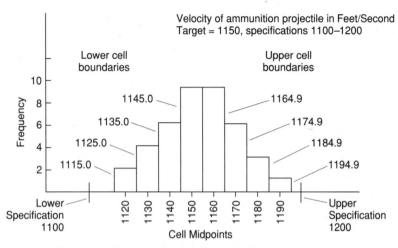

Figure 3.6 Elements of histogram (projectile example).

Calculations for average and standard deviation of grouped data. Now that we have constructed a histogram for the velocity of ammunition projectiles, we show how to calculate the average and the standard deviation for grouped data. This method should be used if there are so many readings that it is not practical to key them all into a calculator or computer.

Lower boundary	Upper midpoint	Cell	Frequency
1115.0	1124.9	1120	2
1125.0	1134.9	1130	4
1135.0	1144.9	1140	6
1145.0	1154.9	1150	9
1155.0	1164.9	1160	9
1165.0	1174.9	1170	6
1175.0	1184.9	1180	3
1185.0	1194.9	1190	1

The first step is to estimate the cell where the average is located. In this example we assume that the actual average is in the cell that has a midpoint of 1160. We choose this cell because it has a high frequency of measurements and is near the middle of the histogram. Three new columns will be added to include

d = number of cells from estimate average
f = frequency of measurements in cell multiplied by number of cells from estimate average
fd^2 = frequency of measurements in cell multiplied by square of cell deviation value

Cell midpoint	Frequency f	Deviation from estimate d	fd	fd^2
1120	2	−4	−8	32
1130	4	−3	−12	36
1140	6	−2	−12	24
1150	9	−1	−9	9
1160	9	0	0	0
1170	6	+1	+6	6
1180	3	+2	+6	12
1190	1	+3	+3	9

$$n = 40 \qquad \sum fd = -26 \qquad \sum fd^2 = 128$$

$$\overline{X} = X_{est} + i \quad = 1160 + 10\left(\frac{-26}{40}\right) = 1160 - 6.5 = 1153.5$$

$$\sigma_n = i\sqrt{\phantom{\frac{128}{40}} - \left(\phantom{\frac{-26}{40}}\right)^2} = 10\sqrt{\frac{128}{40} - \left(\frac{-26}{40}\right)^2}$$

$$= 10\sqrt{3.2 - 0.423} = 16.7$$

Figure 3.7 shows the normal curve (not the histogram) of the velocity of the projectiles, based upon an average of 1153.5 and a standard deviation of 16.7 ft/s.

Looking at Fig. 3.7, we can see that the specification limits of 1100 and 1200 are very close to 3 standard deviations from the center of the curve, so there is a small percentage of product that is outside the specifications. By using the Z formula and the Z table, we determine the proportion of product outside the specifications.

We must calculate the Z values for both specifications and then look up those values in the Z table at the end of the chapter.

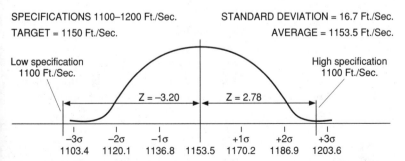

Figure 3.7 Assumed distribution of projectile velocity based upon calculated average and standard deviation.

$$Z = \frac{X_i - \overline{X}}{\sigma} = \frac{1100 - 1153.5}{16.7} = -3.20$$

The lower specification is back 3.20 standard deviations from the center of the distribution.

$$Z = \frac{X_i - \overline{X}}{\sigma} = \frac{1200 - 1153.5}{16.7} = 2.78$$

The upper specification is up 2.78 standard deviations from the center of the distribution. The next step is to look up the Z values in the Z table. The proportion outside the lower specification is 0.00069 (0.069 percent) out of specification on the low side. The proportion of product outside the upper specification is 0.0027 (0.27 percent) out of specification on the high side. (See Fig. 3.7.) The total proportion of product out of specification is $0.00069 + 0.0027 = 0.0034$ (0.34 percent), or 1 out of every 295.

Many veterans of manufacturing, seasoned with years of experience, would be satisfied with 0.34 percent out of specification. Recall from Chap. 1 the target value principle, which supported the idea that the farther the quality characteristic is from the target, the greater the loss. In some situations, product barely out of specification will function well; but likewise there are situations where product barely in specification does not function properly, and there is financial loss.

> QUESTION: The observed range of the velocity measurements was 69 ft/s. Yet the normal distribution curve predicts that the 6σ value to expect is approximately 100 ft/s. Why is there such a difference?
>
> ANSWER: The range that was observed came after measuring only 40 velocities. The power of these simple statistical tools allows us to see the big picture even though we measured only a small sample of product. There really is no conflict here; the observed range from the rather small sample should be less than the predicted 6σ spread.

3.7 Run Charts

In the 1920s Dr. Walter Shewhart developed a method for monitoring a process to see whether it was changing significantly. One requirement for Shewhart control charts to be meaningful is that the measurements must be plotted in *production-order sequence*. The histogram does not record measurements in production sequence order. The run chart does record the measurements in production sequence.

Not all quality characteristics being measured will yield a distribution having the shape of the normal curve. If the shape of the distribution differs greatly from a normal curve, that is not necessarily an

Introduction to Variation and Statistics   61

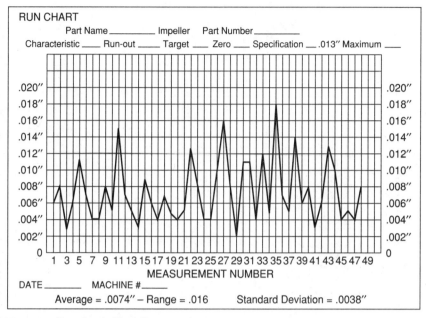

Figure 3.8 Run chart of impeller runout.

indication that something is wrong. Some processes or quality characteristics just naturally form a nonnormal distribution. The most common nonnormal distribution is the skewed distribution. The next example (Fig. 3.8) is a run chart that comes from an inspection check on the runout (eccentricity) of an impeller. The specification limit is 0.013 in *maximum*. This type of characteristic is a "smaller-is-better" type of characteristic. The more runout in the impeller, the greater the loss. The goal in this situation is to find a way to reduce the average runout of the impellers and to reduce the variation from impeller to impeller.

As we visually review the impeller runout measurements on the run chart, we can see that the process looks quite stable, but it does not appear to be symmetric. Remember, the ideal runout is zero. (See the section on control chart pattern analysis for additional information.)

The measurements (in inches) in production sequence are

0.006	0.008	0.003	0.006	0.011	0.007	0.004
0.004	0.008	0.005	0.015	0.007	0.005	0.003
0.009	0.006	0.004	0.007	0.005	0.004	0.005
0.013	0.009	0.004	0.004	0.010	0.016	0.008
0.002	0.011	0.011	0.004	0.012	0.005	0.018
0.007	0.005	0.014	0.006	0.008	0.003	0.006
0.013	0.010	0.004	0.005	0.004	0.009	

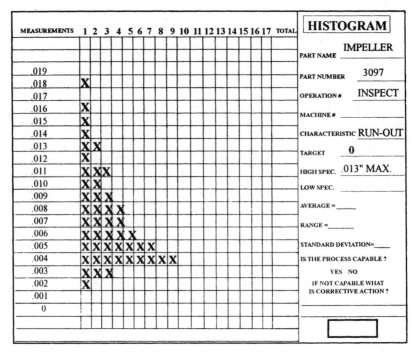

Figure 3.9 Histogram of impeller runout.

Figure 3.9 is a histogram of the impeller runout. The histogram shows that the shape of this distribution differs greatly from the shape of the normal distribution. This shape is referred to as a *skewed* distribution. As with the previous example, in one method of data analysis we calculate the average and the standard deviation of the measurements.

Average = 0.0074 in

Standard deviation = 0.0038 in

Range = 0.016 in

If we blindly assume that these readings are the shape of the normal distribution, then we are led to believe that the output is the shape shown in Fig. 3.10. This leads us to believe that some impellers have a negative runout (less than zero), which is not possible. A disagreement like this could mean

1. There is a mistake in the calculations of the average and/or standard deviation.

2. The shape of the distribution is much different from that of the normal distribution.

Introduction to Variation and Statistics   63

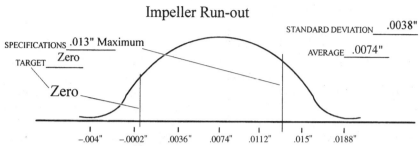

Figure 3.10 Assumed distribution of impeller runout, based upon the calculated average and standard deviation.

3. The process is wildly out of control.

In this case, second reason is the cause of the conflict between the actual and the theoretical. We can see that this is the case when we look at the measurements plotted in a histogram.

Much has been written about the normal distribution. Many computer programs plot out perfectly smooth normal curves over a jagged histogram. The following are just some of the quality characteristics of the skewed distribution rather than the normal distribution: taper, flatness, end-play, out-of-squareness, concentricity, leak rates, and impurity levels. This is a good rule of thumb: When specifications have a maximum limit, the distribution will take the general shape of the skewed distribution rather than the normal distribution.

There are two considerations to note:

1. The standard deviation will not accurately describe a skewed distribution. Recall that earlier we could predict that 68.26 percent of the measurements would be within $\pm 1\sigma$ from the center of the process. If the measurements come from a skewed distribution, the prediction will depart strongly from the actual measurements.

2. The capability calculation (C_p and C_{pk}) will not agree with the actual observations. Chapter 6 will explain the proper way to determine capability for quality characteristics that come from a non-normal distribution.

Do not be too concerned about this. Follow this course of action: Work toward driving the process closer to the target (usually zero). If you are using a control chart to monitor this type of quality characteristic, the strategy is to drive the process out of control on the low side. This would trigger an out-of-control condition on the low side of the control chart. If we are monitoring a "smaller-is-better" type of characteristic, we may actually want the process to go out of control on the low side so that the process is closer to the target of zero. Then new control limits would be calculated for the new improved process.

Note: Do not be too concerned about which shape of distribution describes your processes. Do concentrate on understanding the statistics that describe the distributions. Just as the company is responsible for managing people, machinery, and the methods, the truly advanced company concentrates on managing the variability of the process factors that influence cost, productivity, quality, and customer satisfaction. Once the objective has been identified, we must determine the important factors of the process that are influencing the output. Then the final step is to target those factors at the optimum setting, let the process run, and verify the improvement.

Example 3.4 A refinery is recording the distillation temperature for a petroleum-based product. The specifications are 360 to 380°F. The target temperature is 370°F. The measurements (in degrees Fahrenheit) are

| 366 | 368 | 377 | 371 | 364 | 366 | 373 | 375 | 375 | 378 |
| 369 | 366 | 370 | 374 | 377 | 368 | 366 | 364 | 372 | 375 |

One method of analyzing variable data is to manually enter the measurements into a calculator. We get

Average = 370.7°F   range = 14°F   standard deviation = 4.6°F

Figure 3.11 shows what the assumed distribution looks like, based upon the calculated average and the standard deviation. One problem with this method of analysis is that many people assume that the shape of the distribution that these readings come from is a normal distribution. It would not be correct to determine whether the normal distribution (6σ wide) would fit inside the specification limits of 360 to 380°F.

When we overlay the actual measurements, we see that there is significant conflict between the actual and the theoretical distribution based upon the calculated average and standard deviation.

This distribution is called a *bimodal distribution*. This is an unnatural and unwanted distribution. There are actually two distributions that are the general shape of the normal distribution. Recall that the target temperature was 370°F and the calculated average was 370.7°F. We are lulled into a false sense of security because the target and the average are so close to each other. When we look at Fig. 3.11, we see that there are few readings close to the target and many readings that are quite a distance from the target.

Investigation revealed that two 12-hour shifts operated the refinery unit. The lead operators from the two shifts had never heard of the idea of target value before. Their previous training instructed them to keep the distillation temperature between 360 and 380°F. The process engineering department knew that there were benefits to keeping the temperature close to the target of 370°F. Unfortunately, this information was never passed down to the operations people at the refinery unit. To make things worse, one of the lead operators believed that the unit ran better if the temperature were closer to the low end of the specification. The lead operator on the other shift somehow was under the impression that the temperature should be kept at the higher end of the specification.

Introduction to Variation and Statistics   65

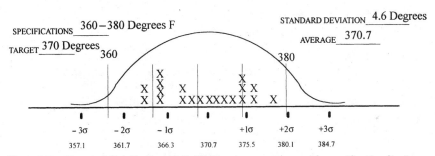

Figure 3.11 Assumed distribution of distillation temperatures with specification limits.

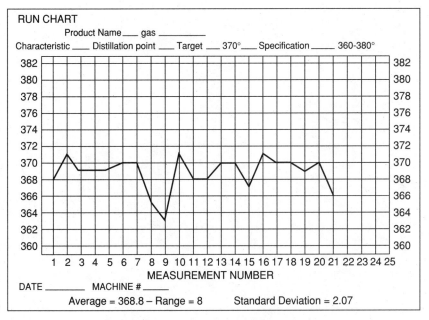

Figure 3.12 Run chart of distillation temperatures after process improvement.

This condition of the two shifts having different targets caused regular problems at this stage in the refinery process and at later stages of the process. Once the situation was explained to the proper people and they agreed to target the process closer to 370°F, improvements in product uniformity were observed and the proportion of off-test product came down.

Figure 3.12 shows the temperatures approximately 60 days later. The average was 368.8°F. The standard deviation was 2.07°F. The range was 8°F, and the distribution was closer to the shape of the normal distribution. There was no need to invest in capital equipment or the modification of existing equipment.

3.8 Using Normal Probability Paper to Evaluate a Nonnormal Distribution

If the distribution of individual measurements differs greatly from the shape of the normal distribution, then calculation of the standard deviation will not give much useful information and could be misleading. One way to evaluate variation and capability for a nonnormal distribution is to use *normal probability paper* (NPP). Figure 3.13 is

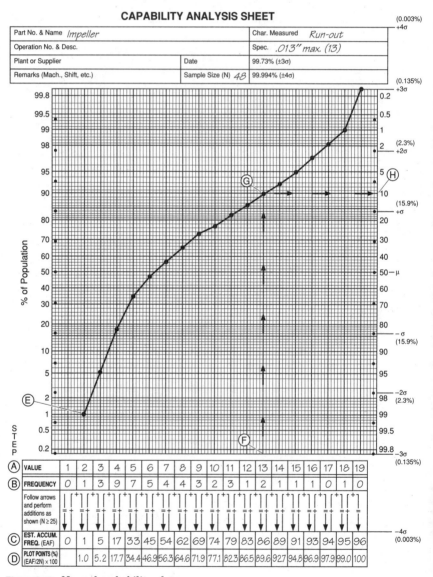

Figure 3.13 Normal probability plot.

an example of NPP. Do as follows to complete the form. [This example will use the impeller runout measurements (Fig. 3.9).]

1. Fill out the header information. The measurements are coded (0.008 in = 8).
2. At the bottom of the form, fill in the values. (location A)
3. Complete the frequency row of the form. (location B)
4. Calculate the estimated accumulated frequency (EAF). (location C)
5. Calculate the plot points in percentage. (location D)

Now a series of points must be plotted on the grid. The first plot point in percentage is 1.0; that percentage is associated with a value of 2 (0.002 in) runout. Read the vertical axis on the left side of the paper, and plot a large dot at the 1 percent location (location E). All the other plot points must be plotted. Connect the dots with a solid line. If the distribution is the shape of a perfect normal distribution, the line will be straight. Note that the line is curved, which indicates that the distribution is skewed.

Using this method, we can accurately estimate the percentage of product that is beyond the specification of 0.013 in (13) maximum. Since we want to determine the percentage of product over 13, the process is as follows:

1. Locate 13 in the row labeled *value*. (location F)
2. Move straight up until the line drawn in is reached. (location G)
3. Move directly to the right and read the percentage (in this case, 10 percent of the product is beyond the specification limit). (location H)

This technique can be used to determine the capability of a process having an output that is not the shape of the normal distribution.

3.9 Taguchi Loss Function

In this section we explain the Taguchi loss function and its application. In the following chapters many methods of evaluating a manufacturing process will be discussed, including

1. Evaluation for statistical control (stability)
2. Evaluation for process capability

The Taguchi loss function method will be used also. The important point is to perform the calculations on the current process, this will be a baseline, and to calculate the loss at different stages of the process improvement effort. Do not be overly concerned if the dollars are not exactly in agreement with the traditional methods of measuring the cost of poor quality; rather, look at the improvement.

For decades managers, engineers, and corporate executives have had limited success with quality improvement efforts because of their attempt to make product within the specification limits. Few people understood that many times there is a need to go beyond just meeting the specification limits.

In the 1960s, when I first went to work in a factory, I was trained by many skilled craftspeople. One of the main rules heard day after day was, "Make the product inside the engineering specification limits." At first glance this is a noble goal and in many instances a real challenge for one reason or another. The main shortcoming is that as long as a product is anywhere between the upper and lower specifications, it is considered acceptable. There is the misconception that no benefit accrues from improving the quality level past that of meeting specifications.

In the meantime, our Japanese counterparts were putting great effort in *advanced quality planning activities*. The effort in Japan was aimed not primarily at just meeting specifications but rather at *learning how to set the processes on target, making product on target, and then reducing variation around the target until the customer was delighted*. Figure 1.6 shows the difference between these two approaches.

We have been led to believe that conformance to specifications means customer satisfaction. U.S. industry is finding out that often this is far from true.

In calculations of the loss function, often Dr. Taguchi uses the *consumer's tolerance* LD_{50} instead of the tolerance limits. This is the point at which 50 percent of the customers view the product as nonfunctional and will take action. The final consumer cares more that the product functions properly at low cost than that the product is in tolerance or out of tolerance. The final customer just wants a product to satisfy her or his wants. A logical approach to increase customer satisfaction is to work on setting our processes and making our products on target and reducing variation around the target.

Figure 3.14 shows four different factories producing the same product with the same tolerance and the loss function calculations for them.

Do not be concerned about what the product is or what characteristic is being measured. The important point is to notice the relative difference in the distribution of product and the corresponding calculated loss. The product from factory 1 is well centered on the target of 20. The distribution of product is barely within the specifications of 16 to 24. Based upon the calculations, the loss per unit is $3.55 in scrap, rework, assembly problems, test failures, customer complaints, and/or the long-term effects of lost business.

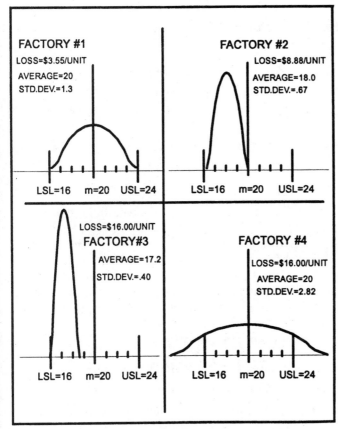

Figure 3.14 Financial loss (Dr. Taguchi's viewpoint). (*Reprinted with permission of the American Supplier Institute, Inc.*)

The product from factory 2 is running on the low side of the target. The good thing is that the product is consistent and that the variation is small (tight normal curve of product). Notice that there is no product on target. According to Taguchi's loss function calculation, the loss per unit is $8.88.

The product from factory 3 is even more consistent than that of factory 2, but it is the most off target of the four factories. The process being off target is the cause of the problem. Although there is very little product out of specification, the calculated loss is $16 per unit.

Finally, the product from factory 4 is centered on the target, but the variation is quite large. There is some product outside the specification limits. The calculated loss from this operation is the same as that for factory 3: $16 per unit.

Types of characteristics

There are four different situations (types of characteristics) that relate to the loss function.

1. *Smaller is better* (SIB)—wear rate, out-of-roundness, contamination level, porosity
2. *Nominal is best* (NIB)—a dimension, mold temperature, bolt torque
3. *Larger is better* (LIB)—tensile strength, fuel efficiency, weld strength
4. *Dynamic characteristics*—thermostat function within a wide temperature band

In each of these situations, there is a different formula for calculating the loss.

The loss function is a newer, more sophisticated approach in the evaluation of quality. This technique goes beyond the idea of scrap, rework, capability, and the C_p and C_{pk} indices. The loss function can be used by product development, design engineering, manufacturing engineering, production engineering, purchasing, and manufacturing, to name just a few.

Why should we use Taguchi's loss function?

- The loss function evaluates variability and targeting.
- The loss function can aid in setting economic tolerances.
- The loss function gives a continuous measurement of quality.
- The loss function translates variation to the language of management: *money*.

Dr. Taguchi's idea of quality is that a lack of quality should be considered a *loss to society*. This idea goes beyond the idea of scrap and rework. Dr. Taguchi's loss function considers returned product, warranty costs, customer complaints, time and money spent by the customer, and eventual long-term loss of market share. Many people say that these factors which Dr. Taguchi takes into account cannot be truly quantified, but everyone agrees that they have a long-term negative effect on the competitiveness of the company.

The following formula for loss L is used when the quality characteristic is NIB for one unit of production only:

$$L(y) = k(y - m)^2$$

where y = value of part or process characteristic being measured
k = financial constant for this particular situation
m = target value of characteristic y

Let us assume that we are interested in the torque of a bolt used in some assembly. The torque that best satisfies the customer's needs is 50 ft · lb. When there is a failure of the product in the field, the cost to repair or replace is $100. The engineering department has estimated that the average customer will be dissatisfied if the torque is 15 ft · lb off target. *Note:* We should not concern ourselves with the exact engineering specification limits. Here $LD_{50} = 15$.

First, we calculate the financial constant k used in the loss function:

$$k = \frac{\text{cost to repair}}{(LD_{50})^2}$$

$$= \frac{\$100}{15^2}$$

$$= \frac{\$100}{225}$$

$$= \$0.45$$

Let us assume that a customer happens to purchase a unit that has a torque value of 46 ft · lb. *Note:* The assumption is that there is an equal cost penalty regardless of the direction in which the reading is off target. We are given

$$m = \text{target} = 50 \text{ ft} \cdot \text{lb} \qquad y = \text{individual reading} = 46 \text{ ft} \cdot \text{lb}$$

$$k = \text{financial constant} = \$0.45$$

Using Dr. Taguchi's loss function, we have

$$L(y) = k(y - m)^2$$

$$L(46) = \$0.45(46 - 50)^2 = \$7.20$$

Thus, the particular customer who purchases a unit that has a torque value of 46 ft · lb will suffer a $7.20 *loss*. Let us now consider the plight of some poor chap who happens to purchase a unit that has a torque value of 41 ft · lb. The loss for this customer is

$$L(y) = k(y - m)^2$$

$$L(41) = \$0.45(41 - 50)^2$$

$$= \$36.45$$

Thus far, we have used Taguchi's loss function to evaluate a unit of product. To get a better view of the process, we should evaluate the entire output of the process. We must look at both targeting and variability. The expanded version of the loss function calculation is

$$L = k[(\bar{y} - m)^2 + \sigma^2]$$

m = target $\bar{y}$ = average of output σ = standard deviation of output

As can be seen, there are three main elements of this formula:

- The financial constant k
- The impact of the process not being on target $\bar{y} - m$
- The impact of variability on the process, σ^2

Here is an example of how to evaluate the entire output of the process. Again, this example is a situation where nominal is best and there is an equal penalty for being off target in either direction. The characteristic being evaluated is the carbon content of a particular iron. We are given

Target value = 2.40%   consumer's tolerance LD_{50} = +0.25%

cost to replace or repair = $4

$$k = \frac{\text{cost to replace}}{(LD_{50})^2} = \frac{\$4}{0.25^2} = \$64$$

We are also given

$\bar{y}$ = average = 2.32% m = target = 2.40%

standard deviation = σ = 0.05%

Now the complete loss function formula for the entire output of the process is

$$L = k[(\bar{y} - m)^2 + \sigma^2] = \$64[(2.32 - 2.40)^2 + 0.05^2] = \$0.57$$

This means that the *loss* for this process is $0.57 per unit in customer dissatisfaction, returned goods, and long-term loss of the market share. If we look at the formula closely, clearly much can be gained by centering the process closer to the target value (2.40 percent).

Example 3.5 The moisture content of paper produced at a paper mill has been causing processing problems at the box plant in the corrugating department. The engineering department has determined that problems are encountered when the moisture content goes below 4.5 or above 7.5 percent. The ideal moisture content (target) is 6.0 percent. This type of quality characteristic is referred to as *nominal is best* (NIB). The process currently has an average moisture content of 6.7 percent and a standard deviation of 0.40 percent. Figure 3.15 shows us that the process has the ability to easily meet specifications if properly centered.

Using the Z formula, we can determine that approximately 2.3 percent of the product is out of specification on the high side. The low specification of 4.5 percent is back more than 5 standard deviations. In this type of situation, you do not really sort through the rolls of paper to find paper that is in specification—you

Introduction to Variation and Statistics 73

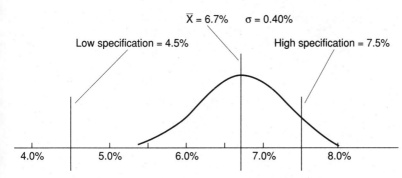

Figure 3.15 Distribution of moisture content (process could meet specification if centered properly).

just try to adjust other settings in the corrugating process to compensate for the variation in moisture content. Great cost is incurred, and productivity suffers.

To evaluate this process in financial terms, we will use the Taguchi loss function. The k factor must be determined first. The cost to repair or replace is $45.00 in this example.

$$k = \frac{\text{cost}}{(\text{tolerance})^2} = \frac{\$45.00}{(1.5\%)^2} = \$20.00$$

Then we plug in the k value to find the loss L:

$$L = k[(\bar{y} - m)^2 + \sigma^2] = \$20.00[(6.7\% - 6.0\%)^2 + (0.40\%)^2] = \$20.00(0.49 + 0.16) = \$13.00$$

Much of the loss from the process would be eliminated if the process could be controlled so that the average would be exactly at the target of 6.0 percent, even with no reduction in variation.

$$L = k[(\bar{y} - m)^2 + \sigma^2 = \$20.00[(6.0\% - 6.0\%)^2 + (0.40\%)^2] = \$20.00(0 + 0.16) = \$3.20$$

The loss has been reduced from $13.00 to $3.20 just by adjusting the process and keeping it centered on the target. There will be some difficulty keeping the process centered on the target, but the financial loss is reduced to about 25 percent of what it was before the adjustment. This should provide motivation for keeping the process centered close to the target of 6.0 percent.

There would be some further reduction in loss if the variation of moisture content could be reduced (the big gains were made by centering the process).

The manufacturer of a new process control system can reduce the variation to the extent that the standard deviation could be held to 0.25 percent or less. The loss will be reduced even more by this reduction in variation.

$$L = k[(\bar{y} - m)^2 + \sigma^2] = \$20.00[(6.0\% - 6.0\%)^2 + (0.25\%)^2] = \$20.00(0 + 0.0625) = \$1.25$$

With this reduction in variation, the loss could be brought down to $1.25. Figure 3.16 shows the three distributions and the loss associated with those distributions.

In later sections of the book, the process capability will calculated and evaluated. The Taguchi loss function will also be determined for the process. This will give the management of the company another method to evaluate a process.

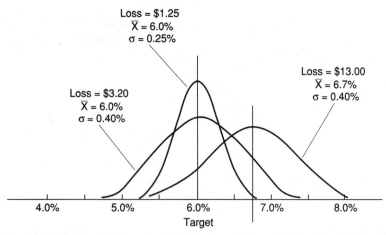

Figure 3.16 Loss calculations for moisture content (three different levels of quality). Moisture content target is 6.0 percent.

Example 3.6: Smaller Is Better. The eccentricity (runout) of a component in an assembly (in thousandths of an inch). We are given

$$\text{Cost to repair or replace} = \$400.00$$
$$\text{Specification} = 6 \text{ maximum} \quad (0.006 \text{ in})$$

The first step is to calculate the k factor for this situation.

$$k = \frac{\text{cost}}{(\text{LD}_{50})^2} = \frac{\$400.00}{6^2} = \$11.11$$

The loss function formula for SIB characteristics is

$$L(y) = k(y)^2$$

The losses associated with six different units produced having eccentricity values of 0.5, 1, 1.5, 2.1, 3.0, and 3.5 are as shown:

$$L(0.5) = \$11.11(0.5^2) = \$2.78$$
$$L(1) = \$11.11(1^2) = \$11.11$$
$$L(1.5) = \$11.11(1.5^2) = \$25.00$$
$$L(2.1) = \$11.11(2.1^2) = \$49.00$$
$$L(3.0) = \$11.11(3.0^2) = \$100.00$$
$$L(3.5) = \$11.11(3.5^2) = \$136.10$$

The loss function based on the previous calculations is shown in Fig. 3.17.

Figure 3.18 graphically shows four examples of Dr. Taguchi's continuous loss function. In example 1, the loss function curve is not symmetric about the target. If the target is 30 ft · lb of torque, we see

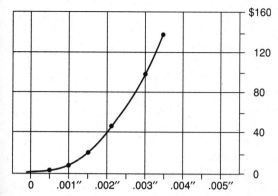

Figure 3.17 Loss function plot (smaller-is-better characteristic).

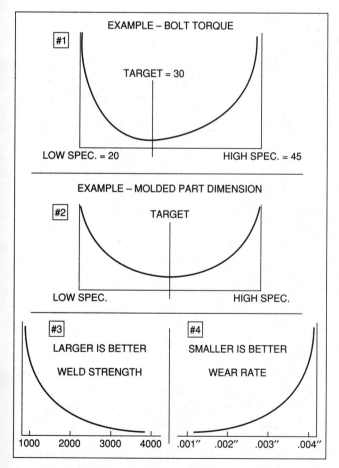

Figure 3.18 Common contours of loss function.

that it is much worse (greater loss) to be *low* off target than *high* off target. In this example, we say we are in deep trouble when the torque gets as low as 20 ft · lb or as high as 45 ft · lb, and the quality is best at the target torque of 30 ft · lb.

Note: In example 1, there would be a different k factor for the two sides of the loss function that would affect the financial penalty.

Example 2 is similar to example 1, except that there is an equal loss for being off target, either high or low. Example 3 applies when the characteristic is *larger is better* (LIB). The larger the reading, the higher the quality level and the customer satisfaction. Example 4 is an SIB type of characteristic. The smaller the value, the higher the level of quality and customer satisfaction.

The loss function formulas are shown for the following types of characteristics:

$L(y) = k(y^2)$ SIB

$L(y) = k\left(\dfrac{1}{y^2}\right)$ LIB

$L(y) = k(y - m)^2$ NIB (one measurement)

$L(y) = k[(\bar{y} - m)^2 + \sigma^2]$ NIB (many measurements)

3.10 Summary

These are the important points to remember from this section:

1. Variable readings from a process, when grouped (summarized), will take definite shapes called distributions.
2. There are two categories of statistics that describe the normal curve:
   a. The first is called the *measure of central tendency*. These statistics answer the question, Where is the center of the process? The statistics that answer that question are the *average, median,* and *mode*.
   b. The second category of statistics is called *measures of dispersion*. Those statistics are the *range* and *standard deviation*.
3. The normal curve is discussed as though it is 6 standard deviations wide: 68 percent of the measurements are between $+1\sigma$ and -1σ, 95 percent of the measurements are between $+2\sigma$ and -2σ, 99.7 percent of the measurements are between $+3\sigma$ and -3σ, and 99.994 percent of the measurements are between $+4\sigma$ and -4σ.
4. We should work toward setting our processes and making our products on target and minimizing the variation of the products around the target.

Introduction to Variation and Statistics 77

5. Variation is the enemy of quality. Variation costs the company money. The smaller that variation is, the smaller the cost penalty to the company. If the variation is quite small but the output of the process is off target a great amount, then the loss will be very great.

3.11 Practice Problems

1. Calculate the average and the median for the following five melt flow index numbers:

   6.1   6.9   6.5   7.2   6.7

2. All characteristics being measured will have the shape of the normal curve, true or false?

3. The smaller the standard deviation is, the narrower the distribution is, true or false?

4. Calculate the average, median, and range of the weight in grams of the following five injection-molded parts:

   522   515   511   508   518

5. The average groove width of a machined forging is 9.08 mm, with a standard deviation of 0.07 mm. Assuming that individual forgings are normally distributed, what percentage of the machined forgings will have a groove width of 9.00 mm or less? What percentage of the castings will have a groove width of 9.20 mm or more?

6. To reduce cost, increase productivity, and improve customer satisfaction, we should concentrate most of our efforts on which of the following?

   a. Processing and shipping as much product as possible every day

   b. Trying to make product in specification

   c. Prioritizing the opportunities for improvement, determining what factors affect the output, then working on getting the process close to the optimum target values

7. The following are temperatures in a leach vessel. The target temperature is 190°F. There are no true specification limits here. Most of the operating personnel say that it is important to keep the temperature within ± 5°F of 190°F. The readings are as follows:

190	190	195	185	190	190	195	190	185
200	191	190	200	192	200	192	195	195
191	191	195	198	198	195	195	195	190
190	190	192	190	191	196	195	198	198
198	198	190	195	194	190	190	191	190

| 195 | 190 | 191 | 190 | 190 | 195 | 190 | 195 | 195 |
| 195 | 195 | 193 | 195 | 195 |

Construct a histogram using these data. Evaluate this situation to determine whether it is satisfactory or unsatisfactory.

8. An injection molding operation is making a part that has an average weight of 282 g and a standard deviation of 2.5 g. What percentage of the product is above 286 g?

9. Sketch a bell-shaped distribution with an average end play $\overline{X}$ of 0.0095 in and a standard deviation σ of 0.0025-in end play.

APPENDIX Z Table

|Z|	x.x0	x.x1	x.x2	x.x3	x.x4	x.x5	x.x6	x.x7	x.x8	x.x9
4.0	.00003									
3.9	.00005	.00005	.00004	.00004	.00004	.00004	.00004	.00004	.00003	.00003
3.8	.00007	.00007	.00007	.00006	.00006	.00006	.00006	.00005	.00005	.00005
3.7	.00011	.00010	.00010	.00010	.00009	.00009	.00008	.00008	.00008	.00008
3.6	.00016	.00015	.00015	.00014	.00014	.00013	.00013	.00012	.00012	.00011
3.5	.00023	.00022	.00022	.00021	.00020	.00019	.00019	.00018	.00017	.00017
3.4	.00034	.00032	.00031	.00030	.00029	.00028	.00027	.00026	.00025	.00024
3.3	.00048	.00047	.00045	.00043	.00042	.00040	.00039	.00038	.00036	.00035
3.2	.00069	.00066	.00064	.00062	.00060	.00058	.00056	.00054	.00052	.00050
3.1	.00097	.00094	.00090	.00087	.00084	.00082	.00079	.00076	.00074	.00071
3.0	.00135	.00131	.00126	.00122	.00118	.00114	.00111	.00107	.00104	.00100
2.9	.0019	.0018	.0018	.0017	.0016	.0016	.0015	.0015	.0014	.0014
2.8	.0026	.0025	.0024	.0023	.0023	.0022	.0021	.0021	.0020	.0019
2.7	.0035	.0034	.0033	.0032	.0031	.0030	.0029	.0028	.0027	.0026
2.6	.0047	.0045	.0044	.0043	.0041	.0040	.0039	.0038	.0037	.0036
2.5	.0062	.0060	.0059	.0057	.0055	.0054	.0052	.0051	.0049	.0048
2.4	.0082	.0080	.0078	.0075	.0073	.0071	.0069	.0068	.0066	.0064
2.3	.0107	.0104	.0102	.0099	.0096	.0094	.0091	.0089	.0087	.0084
2.2	.0139	.0136	.0132	.0129	.0125	.0122	.0119	.0116	.0113	.0110
2.1	.0179	.0174	.0170	.0166	.0162	.0158	.0154	.0150	.0146	.0143
2.0	.0228	.0222	.0217	.0212	.0207	.0202	.0197	.0192	.0188	.0183
1.9	.0287	.0281	.0274	.0268	.0262	.0256	.0250	.0244	.0239	.0233
1.8	.0359	.0351	.0344	.0336	.0329	.0322	.0314	.0307	.0301	.0294
1.7	.0446	.0436	.0427	.0418	.0409	.0401	.0392	.0384	.0375	.0367
1.6	.0548	.0537	.0526	.0516	.0505	.0495	.0485	.0475	.0465	.0455
1.5	.0668	.0655	.0643	.0630	.0618	.0606	.0594	.0582	.0571	.0559
1.4	.0808	.0793	.0778	.0764	.0749	.0735	.0721	.0708	.0694	.0681
1.3	.0968	.0951	.0934	.0918	.0901	.0885	.0869	.0853	.0838	.0823
1.2	.1151	.1131	.1112	.1093	.1075	.1056	.1038	.1020	.1003	.0985
1.1	.1357	.1335	.1314	.1292	.1271	.1251	.1230	.1210	.1190	.1170
1.0	.1587	.1562	.1539	.1515	.1492	.1469	.1446	.1423	.1401	.1379
0.9	.1841	.1814	.1788	.1762	.1736	.1711	.1685	.1660	.1635	.1611
0.8	.2119	.2090	.2061	.2033	.2005	.1977	.1949	.1922	.1894	.1867
0.7	.2420	.2389	.2358	.2327	.2297	.2266	.2236	.2206	.2177	.2148
0.6	.2743	.2709	.2676	.2643	.2611	.2578	.2546	.2514	.2483	.2451
0.5	.3085	.3050	.3015	.2981	.2946	.2912	.2877	.2843	.2810	.2776
0.4	.3446	.3409	.3372	.3336	.3300	.3264	.3228	.3192	.3156	.3121
0.3	.3821	.3783	.3745	.3707	.3669	.3632	.3594	.3557	.3520	.3483
0.2	.4207	.4168	.4129	.4090	.4052	.4013	.3974	.3936	.3897	.3859
0.1	.4602	.4562	.4522	.4483	.4443	.4404	.4364	.4325	.4286	.4247
0.0	.5000	.4960	.4920	.4880	.4840	.4801	.4761	.4721	.4681	.4641

Chapter 4

Measurement System Analysis

4.1 Importance of Measurement System Analysis

Every so often you hear stories like this one.

> When I measure the product or the assembly and the readings are inside the specifications, my boss smiles as though all is well. He says, "Write up the report." But when I go and tell him that some of the readings are out of specification, I am instructed to go out and measure the parts again, measure other products, get a different gage, or have someone else take the measurements.

It is difficult to understand the logic in this example. If we do not want to believe the measurement system when the readings say that we have a problem, then by the same logic we should not believe the measurements when the suspect readings are inside the specification limits.

This can be a disastrous management style. If we continue to use this method of directing the workforce, soon employee morale will suffer, product quality will degrade even more, and the company will become less and less competitive.

Part of a sound process improvement effort should include a formalized procedure for measurement system analysis. The scope of this chapter applies to a manufacturing situation. Many times measurement system analysis is also referred to as a *gage R&R study,* where R&R means *repeatability* and *reproducibility*.

This formalized written method should be part of the "sign-off" process for new gages in the factory. The procedure should be conducted before measurements are taken and before any charting methods are undertaken to evaluate a product or process. The costs incurred to evaluate the measurement system are *preventive* costs and should be assigned to that category when the cost of quality is determined.

In addition to its other benefits, the gage study can tell us if the workforce has been trained adequately regarding the measurement system.

If at the conclusion of the measurement system analysis we find that the results are unsatisfactory and that there is a need to improve the measurement system, then we get together the group of properly trained people and go through the brainstorming exercise, asking the question: What factors in the measurement system affect gage repeatability and reproducibility?

As you may recall, earlier in the book we mentioned that the measurement system is considered part of the process. Therefore, it is logical that we must have a way to evaluate that aspect of the measurement process in a statistical manner. There are two types of errors that can occur when you are using measurement systems:

1. We think there is a problem with the product or process, but in reality there is no problem. The only problem is that the measurement system has given an erroneous false-bad reading. This type of error is referred to as a *type I error* (producer's risk).

2. There is a problem with the product or process, but for some reason the measurement system does not tell us that there is a problem. This type of error is a *type II error* (consumer's risk).

4.2 Definitions

Accuracy is the difference between the observed average of measurements and the true average. To determine the accuracy, it is necessary to obtain precise measurements of sample products. Usually this is accomplished by having the tool room, layout room, or the laboratory measure the product. The true average is determined from these readings. Later this value will be compared to the observed averages from the different people during the repeatability and reproducibility study made on the gage being evaluated. It may not be feasible to measure all the sample parts in the method described previously. An alternative might be chosen. One alternative is: Measure one product at least 10 times by one operator using the gage. Calculate the average of 10 readings, and then compare that value to the true value. The accuracy of the measurement system should be documented and kept on file to be reviewed for unwanted trends.

Precision is the amount of variation observed when one is repeatedly measuring the same product.

Repeatability is the variation in measurements obtained with a gage when it is used several times by one operator while measuring the identical characteristic on the same parts.

Reproducibility is the variation on the average of the measurements made by different operators using the same gage while measuring the identical characteristic on the same parts.

Stability is the difference in the average of at least two sets of measurements obtained with a gage as a result of *time* on the same parts. The stability value should be documented and kept on file for review.

Measurement system error is the combined additive effects of accuracy, repeatability, reproducibility, and stability.

4.3 Steps in Preparing for a Measurement System Analysis

- Define the objective of the study.
- Decide on the numbers of operators, sample parts, and replications.
- Determine the criticality of the characteristic being evaluated.
- Consider the cost of inspection.
- Verify that operators (appraisers) are properly trained.
- Verify that the sample products represent the variability of the process.
- Measurement instruments for variable characteristics must be graduated such that a minimum of 10 increments can be discriminated within the tolerance limits.

4.4 Measurement System Analysis—Long Method

Many times reproducibility is termed *appraiser variation*. It gives indications that the action needed might be as follows:

- Operator training is needed in how to use and/or read the gage.
- Calibrations on the gage could be more clearly defined.

Many times repeatability is referred to as *equipment variation*. It gives an indication that the action needed might be one of these:

- Gage maintenance is required.
- Gage should be redesigned.
- Clamping or locating method of the product could be improved.

84   Chapter Four

Traditionally, problems of gage accuracy and stability and linearity are not as predominant as problems of repeatability and reproducibility. Nonetheless, accuracy, stability, and linearity should be evaluated, documented, and reviewed and action taken when needed.

Example 4.1 A chemical processing plant is concerned about the C_2 content of a product. The specification is 65 to 95. Ten samples are gathered and serialized. Two laboratory technicians will measure the C_2 content of all 10 samples twice, so that the repeatability of the measurement system can be determined. The laboratory technicians must not be biased by knowing the sample number, because they might recall the reading obtained when measuring the sample the first time (see Fig. 4.1).

```
GAGE REPEATABILITY and REPRODUCIBILITY DATA SHEET (Long Method)      C2-CONTENT

-----------------------------------------------------------------------------
         |    1     |    2     |    3     |    4     |    5     |    6     |    7     |    8     |
---------|----------------------------------------------|----------------------------------------------|
Operator | A - (operator 1)                             | B - (operator 2)                             |
---------|----------------------------------------------|----------------------------------------------|
Sample # |1st Trial |2nd Trial |3rd Trial | Range      |1st Trial |2nd Trial |3rd Trial | Range      |
=========|==========|==========|==========|============|==========|==========|==========|============|
    1    |80.3700   |77.1600   |          |3.2100      |79.1800   |78.5100   |          |0.6700      |
    2    |80.1100   |79.5300   |          |0.5800      |78.3300   |79.3200   |          |0.9900      |
    3    |79.6000   |79.3600   |          |0.2400      |80.0000   |78.8200   |          |1.1800      |
    4    |78.6100   |78.6900   |          |0.0800      |78.0400   |81.1000   |          |3.0600      |
    5    |78.1800   |80.7500   |          |2.5700      |78.0900   |78.9400   |          |0.8500      |
    6    |81.2800   |81.3300   |          |0.0500      |78.4600   |81.0900   |          |2.6300      |
    7    |81.3700   |81.3500   |          |0.0200      |81.3100   |81.3900   |          |0.0800      |
    8    |81.1700   |78.9500   |          |2.2200      |81.3100   |81.3000   |          |0.0100      |
    9    |78.4000   |78.5000   |          |0.1000      |80.6000   |81.7600   |          |1.1600      |
   10    |78.5000   |78.5000   |          |0.0000      |81.0300   |78.7800   |          |2.2500      |
   11    |          |          |          |            |          |          |          |            |
   12    |          |          |          |            |          |          |          |            |
   13    |          |          |          |            |          |          |          |            |
   14    |          |          |          |            |          |          |          |            |
   15    |          |          |          |            |          |          |          |            |
=========|==========|==========|==========|============|==========|==========|==========|============|
Totals   |797.5900  |794.1200  |          |9.0700      |796.3500  |801.0100  |          |12.8800     |
---------|----------------------------------------------|----------------------------------------------|
         |--->|797.5900 .|       |0.9070     |  |--->|796.3500 |       |1.2880     |
         |----------|        ------------                  |----------|       -----------
         |          |<----       _                         |          |<----      _
         |----------|            R                         |----------|           R
     Sum |1591.7100|             A                     Sum |1597.3600|            B
         |----------|                                      |----------|
     _                                                   _
     X   |79.5855  |                                    X |79.8680  |
     A   -----------                                    B -----------


_
R   = 0.9070       # Trials    D         (  R  ) x (  D  ) = UCL *       Max. X = 79.8680
A                    4                                 4       R
                   -----------------     (1.0975) x (3.2700) = 3.5888   Min. X = 79.5855
_                                                                       _
R   = 1.2880        2   | 3.27                                          X Diff. = 0.2825
B                       |
Sum = 2.1950
                    3   | 2.58          * Limit of individual R's :
_
R   = 1.0975

Notes:
```

Figure 4.1 Data sheet for measurement system analysis by the long method (C_2 content).

Figure 4.1 is the data sheet for the gage study. The averages for technicians A and B are determined:

$$\overline{X}_A = 79.59 \quad \overline{X}_B = 79.87$$

The ranges and then the average range for the two technicians' readings are calculated:

$$R_A = \frac{9.07}{10} = 0.907 \quad R_B = \frac{12.88}{10} = 1.288$$

One point of concern is that the average range for technician B is more than 40 percent greater than the average range for technician A. This should be noted and investigated later.

The average difference between the two technicians' measurements must be determined. (This value is needed to determine the reproducibility portion of the measurement system evaluation.)

$$\overline{X}_{\text{diff}} = 79.87 - 79.59 = 0.28$$

Figure 4.2 shows the repeatability and reproducibility calculations. The next step is to convert these values to a percentage of the tolerance. The gage study calculations tell us that 16.68 percent of the tolerance is consumed by the repeatability (equipment variation) of the measurement system.

In this example, 0 percent of the tolerance is being consumed by the repeatability (appraiser variation) of the measurement system. The repeatability and reproducibility of the measurement system are 16.68 percent of the tolerance. This would be considered acceptable.

4.5 Gage Correlation

A petroleum refinery is having problems making the in-line analyzer and the laboratory analysis agree. Figure 4.3 shows fifteen samples that were measured both by the in-line analyzer and the laboratory. In the ideal situation, the in-line analyzer and the laboratory would have the exact same reading for the same sample.

The specification for this characteristic is 500 ± 200 ppm. It is obvious that the two measurement systems do not agree with each other very well. The first question is, Which method of measurement should we believe, or are they both wrong? The laboratory has a primary standard to calibrate its equipment with. The records kept by the laboratory show that the laboratory instrument is calibrated properly. The measurements are not in chronological order; they are in ascending order based upon the laboratory measurements. This shows that (1) the in-line analyzer always gives lower readings than the laboratory analysis and (2) when the readings are at the low end of the specification, the two measurement methods give readings very close to each other.

In the long term, the in-line analyzer must be fixed so that people at the production unit can believe the in-line analyzer and not be so dependent upon the laboratory. The in-line analyzer gives almost instan-

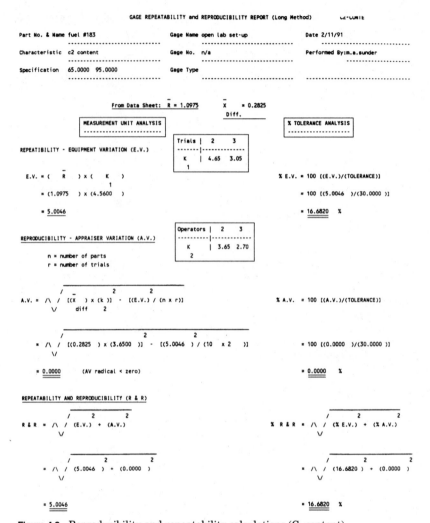

Figure 4.2 Reproducibility and repeatability calculations (C_2 content).

taneous feedback. The laboratory takes from 2 to 6 h for feedback, and costs are much higher.

Figure 4.4 visually shows the extent to which the two measuring instruments correlate. There seems to be a correlation (straight line) between the two methods of measurement. In the short term, it might make sense to see if there is a *compensation factor* that can be used to compensate for the lack of agreement, so that the measurements from the in-line analyzer can be adjusted to determine what the laboratory measurements would be. This is not a good thing to do, but for the

Lab	In-line	Difference
680	610	70
650	585	65
640	585	55
600	550	50
590	540	50
510	465	45
505	450	55
480	445	35
430	410	20
410	385	25
380	365	15
360	350	10
340	330	10
320	320	0
320	315	5

Figure 4.3 Gage correlation data between laboratory measurements and in-line analyzer, in descending order based upon laboratory measurements (ppm contamination characteristic).

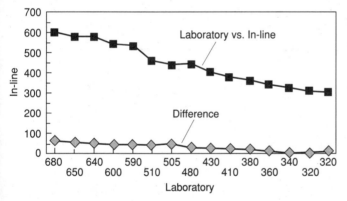

Figure 4.4 Scatter plot showing correlation of laboratory readings versus in-line analyzer and the difference between the two. Measurements in descending order.

short term it is the action to take until the improvements can be made on the in-line analyzer.

The compensation factor can be determined either by reviewing Fig. 4.3 and interpolating the value or by reviewing Fig. 4.4 and reading the difference line to determine the compensation factor. The bottom line of Fig. 4.4 shows the amount that the in-line analyzer is off, depending upon the laboratory reading.

There should be a procedure for verifying measurement system adequacy before the measurement system is released to be used in production. This may cause more work for the department responsible for maintaining instrumentation at the complex, but it will be a good business decision. Investing the time and money to verify the adequacy of the measurement system will prevent problems from arising downstream at a much more costly location in the process. (Remember the leveraged quality principle from Chap. 1.)

4.6 Measurement System Analysis for Attribute Characteristics (Visual Inspection)

Attribute-type measurements are binary—either good or bad. When the quality characteristic is strictly an attribute type of characteristic, the goal is to evaluate the repeatability and reproducibility of the people categorizing the product as good or bad.

Recall the type I and type II errors that can be made:

I. Classifying a product as nonconforming when the product actually does conform to the requirements. This is considered a *false alarm*.

II. Classifying a product as conforming when the product does not actually conform to the requirements. This is considered a *miss*.

The objective of this type of a study is to evaluate the measurement system to

1. Determine the effectiveness of the appraisers doing the inspection. Effectiveness is the appraiser's percentage of making the right decision—accepting good product and rejecting bad product.
2. Determine the probability of a false alarm—rejecting good product.
3. Determine the probability of a miss—not rejecting a bad product.
4. Determine the bias—the ratio of false alarms to misses.

An evaluation of the inspectors (appraisers) can be made concerning effectiveness, false alarms, misses, and bias. Then the company can set guidelines for people to do visual inspections. The analysis will point to the areas needing improvement.

Steps in conducting study (visual inspection)

1. Choose the number of appraisers, of products to be inspected, and of inspections per product. The requirements are shown in the table at the top of page 89.
2. Gather products to be inspected. One-third of the product should be good product, and about one-third should be bad product. The last third should be marginal product. (A person skilled and

Sample Size for Attribute Inspection

Number of appraisers	Minimum number of parts	Minimum number of inspections per part
1	24	5
2	18	4
3 or more	12	3

knowledgeable about the characteristic must determine the classification of the product, good or bad.) Serialize the products so that the appraisers are not biased. In the following example, the "good" parts are numbers 1, 3, 4, 7, 9, and 12. The bad parts are numbers 2, 5, 6, 8, 10, and 11.
3. Have the appraisers inspect the products the required number of times and record the results on the data sheet. See Table 4.1.
4. Complete the data analysis work sheet (Table 4.2). Calculate the following for each of the appraisers:
   a. The number of good correct decisions
   b. The number of bad correct decisions
   c. The number of total correct decisions
   d. The number of false alarms
   e. The number of misses
   f. The grand total

TABLE 4.1 Data Sheet

Product	Classi-fication*	Appraiser A			Appraiser B			Appraiser C		
		1	2	3	1	2	3	1	2	3
1	Good	G	G	G	G	G	B	G	G	G
2	Bad	B	B	B	B	B	B	B	B	B
3	Good	G	G	G	B	B	G	G	G	G
4	Good	G	G	G	G	G	G	G	G	G
5	Bad	B	B	B	B	B	B	B	B	B
6	Bad	G	B	G	B	B	B	B	B	B
7	Good	G	G	G	B	B	G	G	B	G
8	Bad	B	B	B	B	B	B	B	B	B
9	Good	G	G	G	G	G	G	G	G	G
10	Bad	B	B	B	B	G	B	B	B	B
11	Bad	B	B	B	B	B	B	B	B	B
12	Good	G	G	G	G	G	G	G	G	G

*An agreed to classification as to whether the part is good or bad. This evaluation will determine if the appraisers classified a good part as good and a bad part as bad. Example product 1 is a good part. Appraiser B classified the part as bad on the third evaluation. This is an example of a false alarm.

SOURCE: J. L. Hradesky, *Productivity and Quality Improvement*, McGraw-Hill, New York, 1988.

TABLE 4.2 Attribute (Visual) Data Work Sheet

Appraiser	Column 1 Good correct	Column 2 Bad correct	Column 3 Total correct	Column 4 False alarms	Column 5 Misses	Column 6 Grand
A	18	16	34	0	2	36
B	13	17	30	5	1	36
C	17	17	34	1	1	36

TABLE 4.3 Attribute Gage Study Calculations Work Sheet (Visual Inspection)

	Calculations [column values from Table 4.2]			
Appraiser	Effectiveness [3/6]	Probability of false alarm [4/(1 + 4)]	Probability of a miss [5/(2 + 5)]	Bias* B_{FA}/B_M
A	$\dfrac{34}{36} = 0.94$	$\dfrac{0}{18 + 0} = 0.00$	$\dfrac{2}{16 + 2} = 0.11$	$\dfrac{0}{0.1872} = 0$
B	$\dfrac{30}{36} = 0.83$	$\dfrac{5}{13 + 5} = 0.28$	$\dfrac{1}{17 + 1} = 0.06$	$\dfrac{0.3372}{0.1200} = 2.81$
C	$\dfrac{34}{36} = 0.94$	$\dfrac{1}{17 + 1} = 0.06$	$\dfrac{1}{17 + 1} = 0.06$	$\dfrac{0.1200}{0.1200} = 1.0$

*To determine bias, the probability of a false alarm (FA) and the probability of a miss (M) are first converted to a bias factor (see Table 4.1).

5. Complete the calculation work sheet (Table 4.3) for each of the appraisers. You must refer back to the values in Table 4.2 to determine effectiveness, P_{FA} and P_M. For example, in Table 4.3 to calculate effectiveness the formula is the value from Table 4.2 in column 3 divided by the value in column 6.
   a. Determine the effectiveness.
   b. Determine the probability of a false alarm.
   c. Calculate the probability of a miss.
   d. Determine the bias. (See Tables 4.4 and 4.5 to calculate the bias.)
6. Compare the results with the acceptance criteria and determine whether corrective action is needed. The suggested criteria for acceptance are as follows:
   a. Effectiveness E. An acceptable value for the effectiveness is $E \geq 0.90$. If $0.80 \leq E < 0.90$, it should be considered as marginal. If $E < 0.80$, it is unacceptable.

Measurement System Analysis

TABLE 4.4 Determine Bias Value

Probability of false alarm	Bias of false alarm or miss	Probability of false alarm	Bias of false alarm or miss
0.01	0.0264	0.26	0.3251
0.02	0.0488	0.27	0.3312
0.03	0.0681	0.28	0.3372
0.04	0.0863	0.29	0.3429
0.05	0.1040	0.30	0.3485
0.06	0.1200	0.31	0.3538
0.07	0.1334	0.32	0.3572
0.08	0.1497	0.33	0.3621
0.09	0.1626	0.34	0.3668
0.10	0.1758	0.35	0.3712
0.11	0.1872	0.36	0.3739
0.12	0.1989	0.37	0.3778
0.13	0.2107	0.38	0.3814
0.14	0.2227	0.39	0.3836
0.15	0.2323	0.40	0.3867
0.16	0.2444	0.41	0.3885
0.17	0.2541	0.42	0.3910
0.18	0.2613	0.43	0.3925
0.19	0.2709	0.44	0.3945
0.20	0.2803	0.45	0.3961
0.21	0.2874	0.46	0.3970
0.22	0.2966	0.47	0.3977
0.23	0.3034	0.48	0.3984
0.24	0.3101	0.49	0.3989
0.25	0.3187	0.50	0.3989

TABLE 4.5 Special Cases in Computing Bias

Probability of a false alarm P_{FA}	Probability of a miss P_M	Bias	Decision or action
0	>0	0	Unacceptable
>0	0	No value	Use E, P_{FA}, and P_M directly
0	0	No value	Same as $B = 1$, acceptable
>0.5	≤0.5	>1.5	Unacceptable
≤0.5	>0.5	<0.5	Unacceptable
>0.5	>0.5	No value	Unacceptable, because P_M and P_{FA} are more than 0.5—bias is not important

b. Probability of a false alarm P_{FA}. A value of 0.05 or less is considered acceptable; $0.05 \leq P_{FA} \leq 0.10$ is considered marginal; and $P_{FA} > 0.10$ is considered unacceptable.
c. Probability of missing a bad product P_M. A value $P_m < 0.02$ is acceptable; $0.02 < P_M \leq 0.05$ is considered marginal; and $P_M > 0.05$ is considered unacceptable.
d. Bias B. A value $0.80 < B < 1.20$ is considered acceptable; $0.50 < B < 0.79$ or $1.2 < B < 1.5$ is considered marginal; and $B < 0.5$ or $B > 1.5$ is considered unacceptable.

Conclusion

The conclusions concerning the three appraisers are as follows:

Appraiser A:

Effectiveness of 0.94 is acceptable.

Probability of false alarm of 0 is acceptable.

Probability of a miss of 0.11 is unacceptable.

Bias value of 0 is unacceptable.

Appraiser A is biased toward missing bad products.

Appraiser B:

Effectiveness of 0.83 is marginal.

Probability of false alarm of 0.28 is unacceptable.

Probability of a miss of 0.06 is unacceptable.

Bias value of 2.81 is unacceptable.

Appraiser B is biased very much toward missing bad products.

Appraiser C:

Effectiveness of 0.94 is acceptable.

Probability of false alarm of 0.06 is marginal.

Probability of a miss of 0.06 is unacceptable.

Bias value of 1.0 is acceptable.

Appraiser C is doing the best of all three appraisers.

All three appraisers are in need of reducing the frequency of missing bad product. This is a type II error. Somehow the management must communicate to the appraisers that they should be more critical in doing the visual inspection.

Summary

There should be a written procedure that the company follows which will determine the adequacy of the measurement system. This activity would fall into the prevention area in the cost-of-poor-quality report. If the measurement system is not acceptable, corrective action should be initiated to improve it.

Chapter

5

Shewhart Control Charts

In Chap. 1, we discussed the development of an "early warning" or prevention system. Such a prevention system means that an "alarm" will go off before "bad things" happen. An alarm would be an out-of-control condition. "Bad things" would mean a product that is out of specification or, in some instances, a product that meets specification but does not function properly. The Shewhart control chart for variables is the most common statistical tool used to help attain this goal.

During the 1920s, Dr. Shewhart, a statistician working for AT&T, was trying to develop a statistical tool for use on the production floor to get the immediate attention of the machine operator, supervisor, setup operator, or other appropriate person when the process had "changed noticeably." *Changing noticeably* means that a process is out of control. Then the person responsible for monitoring the process would investigate the cause of the change and take appropriate action.

Control charts can be used to monitor the quality characteristic of a product or a process parameter. Companies with well-established systems of process improvement usually try to monitor the process at regular intervals and take action when the control chart indicates action is needed. The tools used by such companies are known as *control charts for variables,* and they are simple, yet powerful tools that aim for a *never-ending improvement* of the processes. Some of the more common types of control charts used today are:

1. Average and range charts ($\overline{X}$ and R)
2. Median and range charts ($\widetilde{X}$ and R)
3. Average and standard deviation charts ($\overline{X}$ and σ)
4. Individual and moving range charts (X and R_m)
5. Moving average and moving range charts ($\overline{X}_m$ and R_m)

6. Multistream charts
7. Nominal charts (low-volume, small-lot applications)
8. Modified-limits control charts

5.1 Variables Chart

This chapter will deal with variable measurements. As you remember, variable-type readings are ones that have come from a continuous distribution. Examples of variable-type readings are a temperature in degrees Fahrenheit, the chemical composition of a material in percentage, a pressure setting on a machine, the hardness value of a material, a dimension of a machined or molded part, the tensile strength of a material, the pressure of a hydraulic system, the viscosity of a material, and an electrical characteristic of a circuit board.

Wherever there is a need to work on improving our processes so that they are running as consistently as possible, then we should consider using Shewhart control charts to help us reach this goal. The objective, when you are using control charts, is to improve the processes to the extent that they will be operating in a state of perfect statistical control and will be closely centered on the target. When this has been attained, there is only *common-cause* variation influencing the process. If a process is in a state of statistical control, centered close to the customer's target, and can easily meet the specification requirements, probably there is an early-warning system in place. Thus there is a strong likelihood that the alarm will go off, signaling that the process has gone out of control. But bad product will not be produced, and the process will not fluctuate to the extent that any significant loss occurs. Admittedly, in some instances the process goes out of statistical control, and out-of-specification product is produced. This is an undesired situation.

Note: Some processes that are in statistical control and are targeted very well are *not capable* of meeting specifications. In other words, common-cause variation is causing undesirable things to happen (scrap rework, downgrade of product, customer returns or complaints, low production rates, or high costs). The *management* of the company must change the process for the better so that common-cause variation is reduced.

Once we have identified a particular operation or product characteristic as being vitally important to internal operations or external customer satisfaction, we may choose to monitor that product or process characteristic by using a control chart. The concept of using control charts is related to the idea of evaluating sample statistics compared to calculated control limits. (Sample statistics are averages and ranges, or median and ranges, or individual and moving ranges, or moving averages and moving ranges, or averages and standard deviations, depending on what type of variable control chart is being used.) At first glance, this idea may appear to be illogical. But if we want to find out very early if a process is changing significantly, we should be monitoring the sample statistics calculated from the individual measurements to see if the process is staying within its own natural limits. Sample statistics are much more sensitive to change than are individual measurements.

5.2 Control versus Capability

Much confusion occurs when discussing the two topics of statistical control and capability. The confusion between *control* and *capability* can be minimized if we remember a few basic guidelines:

1. Blueprint specifications or tolerances are for individual measurements (X's).
2. Statistical control limits are for sample statistics (averages and ranges, medians and ranges, or individual and moving ranges).
3. The term *capability* means the ability to consistently meet specifications. (Will the distribution of individuals fit between the specifications?) Companies should concentrate most of their efforts on statistical control and targeting the processes, rather than just making product in specification.
4. The term *statistical control* means that the sample statistics are consistently within the calculated control limits and that there is no unnatural pattern of variation.

Once a process has been improved to the extent that it is in control and on target, that is as good as *that* process will ever be.

Note: If you modify the process by altering the material, speeds, feeds, subcomponent part design, temperature, chemical composition, cycle time, etc., then this becomes a new process. Control and the capability will be different, hopefully improved.

It is important to understand that there are two types of problems in most manufacturing operations:

1. Problems of statistical control (special-cause variation)
2. Problems of capability (common-cause variation)

If a process is out of control (sample statistics outside the calculated control limits or other nonrandom patterns of variation exist), then there is something to investigate and correct with the process so that it will be even more improved. An out-of-control condition means that special-cause variation is influencing the process or product. This is an undesirable situation. It is usually economically advisable to eliminate the source of special-cause variation.

If a process or product characteristic is not satisfactory, then the process is not able to easily meet specifications and expected levels of customer satisfaction. (Often this is referred to as a process that is *not capable.*) Product is being produced that must be scraped, reworked, or downgraded on a regular basis, or the process is acting up in such an unsatisfactory fashion that it is causing problems with product at the next stage of processing.

Manufacturing operations have some processes running in each of these four situations this very minute. The four situations are shown graphically in Fig. 5.1.

1. In control and capable
2. In control but not capable
3. Out of control but meeting specification requirements
4. Out of control and making out-of-specification product

Process 1 should not be touched. Let this process continue to make product. Process 1 is running as well as it can and is making satisfactory product. Satisfactory product means product well within the specifications and providing high levels of customer satisfaction.

In process 2, there is nothing to fix or improve with the existing process. The process is in control, as it exists today. This process is stable, consistent, repeatable, and predictable, *but unfortunately this process is making some bad product (is not capable).* This is a big opportunity for improvement.

We still have several options for process 2: We could change process factors one at a time, hoping for improvement; optimize the process by using the design of the method; or open up the specification to accommodate what the process needs. You may ask, What does the process need? The answer varies from case to case, but the general answer is that the process needs specification limits that are a minimum of 10 standard deviations away from each other (or $\pm 5\sigma$ from the center of the process). There are some companies that will say $\pm 4\sigma$ instead. (This will be explained in greater detail in Chap. 6.)

Caution! Opening up specifications does nothing to improve the process, product, or customer satisfaction. On one hand, opening up the specifications will reduce the internal rejection rate; but on the other hand, product that was being rejected will now be used in the field. There is the chance that customer satisfaction will suffer as a result of opening up the specifications.

In process 3, something is acting up in the process. (It is out of control.) Special-cause variation is influencing the process. The process currently is making satisfactory product (well within specification). But the process is acting up. There is something to investigate and

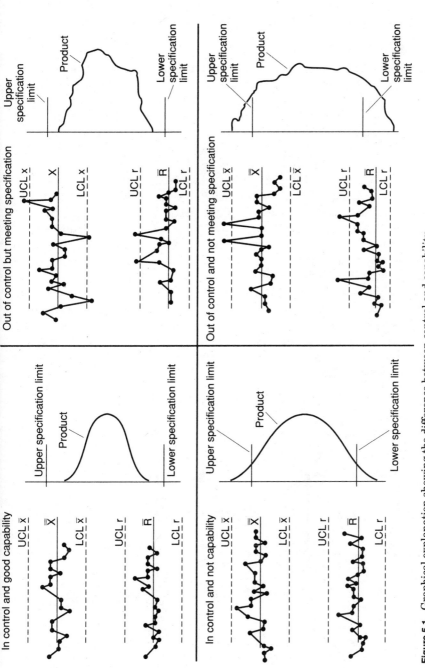

Figure 5.1 Graphical explanation showing the difference between control and capability.

correct. Yet, it is not acting up so badly that it is making out-of-specification product. Nor are the process parameters causing problems with other downstream product characteristics or process parameters.

Process 4 is acting up. (It is out of control.) Special-cause variation is influencing the output. The process is not stable, consistent, repeatable, or predictable. And it is making out-of-specification product.

In summary, *processes 2 and 4 are the two biggest opportunities for improvement in productivity, cost reduction, quality, and customer satisfaction.* They must be approached differently. Each is a different type of problem. Process 2 is exhibiting a problem of capability. *The common-cause variation inherent in the process is so large that it is causing an unsatisfactory condition.*

Process 4 is not attaining statistical control. *Special-cause variation is creating an unsatisfactory condition.* There are two things to do: (1) Work on finding the causes of special-cause variation, and eliminate them in the future. (2) Once the process has been brought into statistical control, evaluate the capability of the process; perhaps the capability will be acceptable. If the capability is not acceptable, then action must be taken to reduce common-cause variation.

Thus far we have not addressed the topic of customer satisfaction. Figure 5.2 is another 2 × 2 × 2 matrix similar to the one seen in Chap. 1. Most of the printed literature stresses the topics of statistical control and capability. In reality, we are concerned mainly with high levels of customer satisfaction. If there is a need to improve the levels of

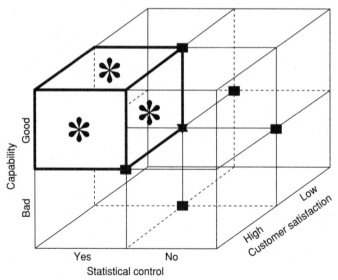

Figure 5.2 Possible combinations of control, capability, and customer satisfaction.

customer satisfaction, usually the means to do so involve work on improving both the statistical control and the capability of the process.
The table of the eight processes shows the possible combinations.

Process no.	High customer satisfaction	Statistical control	Good capability
1	Yes	Yes	Yes
2	Yes	Yes	No
3	Yes	No	No
4	Yes	No	Yes
5	No	Yes	Yes
6	No	Yes	No
7	No	No	Yes
8	No	No	No

Earlier Fig. 5.1 graphically displayed the combinations of statistical control and capability. We should think of the products from the first four processes as being in high demand by the consumer and experiencing few, if any, field failures or customer complaints. The products from the last four processes sit on the shelf, and few are sold; of those sold, there are many field failures and/or customer complaints.

Process 3 with high customer satisfaction, being out of control, and having bad capability does seem to be a possible contradiction.

5.3 Identifying Opportunities for Improvement

The *process* causes variation. The elements of the process are

1. Material
2. Machine and tooling
3. People
4. Method
5. Miscellaneous (environment, measurement system, and the product design itself)

Remember, *statistical problem solving* (SPS) stresses the importance of concentrating on improving the *process*. Then the product will improve.

Types of variation

There are two types of variation. The first is *common-cause variation,* sometimes referred to as "chance-cause" variation. It is always present, is small in magnitude, and will generate an in-control process. It represents the true process. If this type of variation is causing an unsatisfactory situation, it is the responsibility of the management of the company to minimize it. The second type of variation is called *special-cause variation,* sometimes referred to as *assignable-cause variation.* Special-cause variation occurs mysteriously. It is large in magnitude and will trigger an out-of-control condition. It is not part of the true intended process. If this type of variation is causing an unsatisfactory situation with the process, then it is more the responsibility of the local workers to find the cause and try to prevent it from influencing the process in the future. If local workers do not have the authority to make the necessary changes, then the management of the company must take action.

Power of the control chart

The Shewhart control chart will sort out common-cause variation from special-cause variation. Then we can determine whether the local workforce is responsible for eliminating special-cause variation or the management team should work on reducing common-cause variation.

Goals in the use of control charts

In most situations, when using control charts as a tool to aid in process improvement, you have three goals:

1. Find the cause of and eliminate special-cause variation.
2. Adjust the process so that it is centered close to the target.
3. Calculate the capability of the process once it is in a state of statistical control. If the capability is not acceptable, then reduce common-cause variation until the capability is acceptable.

When using control charts for variables, you have three goals. Work on improving the process until

1. The process is in control.
2. The C_p value is 1.67 minimum.
3. The C_{pk} value is 1.67 minimum.

Note: The second and third goals relate to the subject of capability and are covered in Chap. 6. Another way of saying this is: Improve the critical processes so that we are operating in a state of statistical control and the output is considered satisfactory by all involved.

5.4 Preplanning Checklist

There are numerous considerations that you must take into account when preparing to use control charts. Here is a checklist of some planning items to consider:

1. Has the Pareto principle been used to identify the big opportunities for improvement in scrap, rework, non-value-added effort, customer complaints, field failures, etc.?
2. Have brainstorming sessions been held with a cross-functional team to identify customer wants, important product performance characteristics, product quality characteristics, and process parameters that affect product performance and customer satisfaction? Is the number of characteristics to be charted manageable?
3. Are the measurements variable in nature or attribute? Has a measurement system analysis been conducted?
4. Have all the parties involved agreed to the sampling frequency and sample size?
5. Have the production supervisor and factory workers had input to these items, and have basic contingency plans been written and distributed to all parties involved?
6. Have the target values for the characteristics to be monitored been determined?

5.5 Case Study

A manufacturer of wood furniture is in the early stages of using statistical tools to help improve processes and products. The company has chosen to use a control chart on one of its processes that is running quite well. The team thinks this will not be an opportunity for great improvement, but the company wants to start out with a process that is running well. Then it will tackle more difficult processes. The company has a written internal policy that specifies the minimum goals for capability:

- Statistical control for important characteristics
- A C_p value of 1.33 minimum
- A C_{pk} value of 1.33 minimum

This example will evaluate the thickness of a planed board. The operation runs two 8-h shifts. The specification is $\frac{1}{2}" + \frac{1}{32}" - 0$. In the past, a 6-in hook rule graduated in $\frac{1}{64}$- or $\frac{1}{100}$-in increments was used. There was an agreement to use a dial caliper at least during the learning phase of this exercise. Both shifts said that the job should run on the high end of the specification. The engineering department said that it was equally bad if the board was too thick or too thin. This leads us to believe that the target should be the midpoint of the specification, or $^{33}/_{64}$ in (0.515 in). The thickness measurement is fast and inexpensive. The team agreed to measure five consecutive boards every 45 min. The manufacturing department will run the process just as it has in the past. Any change or adjustment to the process will be recorded. After 25 samples the control limits will be calculated to determine if the process was in control and targeted well. If the process shows a good state of control, then the capability of the process will be evaluated. The measurements on the control chart are coded from a reference of 0 = 0.500 in: Example +17 = 0.517 in. Any individual measurement between −0 and +30 is within the specifications. Figure 5.3 is the control chart where the individual measurement is recorded along with the averages and ranges and plotted along with the calculated control limits.

At first glance at the individual measurements and the averages, it seems that the process is aimed closer to +20 (0.520 in) than the agreed-to target of 0.515 in. Now the *upper control limit* (UCL) and *lower control limit* (LCL) will be calculated to determine how stable the process is. There is a comment on the chart that the cutter head was changed after sample 18. All 25 samples will be used to determine the control limits. If sample 18 or 19 is outside the control limits or if something else is unusual about those samples, then a

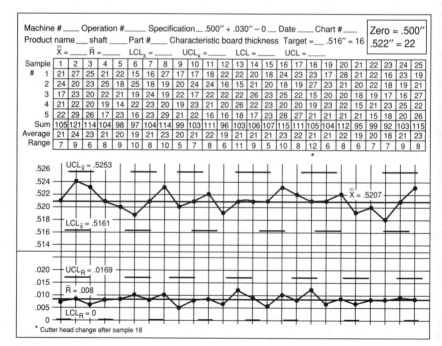

Figure 5.3 Average and range charts with points plotted and calculated limits.

decision will be made concerning whether any readings should be discarded.

$$\bar{\bar{X}} = \frac{\Sigma \bar{X}}{k} = \frac{518}{25} = 20.7 = 0.5207 \text{ in}$$

$$\bar{R} = \frac{\Sigma R}{k} = \frac{199}{25} = 7.96 = 0.0080 \text{ in}$$

$$\text{UCL}_{\bar{X}} = \bar{\bar{X}} + A_2 \bar{R} = 20.7 + 0.577(8.0) = 25.3 = 0.5253 \text{ in*}$$

$$\text{LCL}_{\bar{X}} = \bar{\bar{X}} - A_2 \bar{R} = 20.7 - 0.577(8.0) = 16.1 = 0.5161 \text{ in}$$

$$\text{UCL}_R = D_4 \bar{R} = 2.115(8.0) = 16.9 = 0.0169 \text{ in}$$

$$\text{LCL}_R = D_3 \bar{R} = 0(8.0) = 0$$

The grand average $\bar{\bar{X}}$ of 0.5207 in confirms our visual analysis that the process is running high relative to the agreed-to target of 0.515 in. The next step is to determine what scale should be used for the average and range chart. There are two main considerations in the selection of the scale for the average and range chart:

1. The scale should be set so that the variation in the process can be easily seen. One reason is that pattern analysis of the completed control chart will be conducted once the averages and ranges have been plotted and the limits added to the control chart.

2. The scale should easily accommodate the extreme observed readings and the control limits. Sometimes when a process is widely out of control, it is difficult to accomplish both these concerns.

Chart analysis

Based upon the first 25 samples, the following statements can be made.

1. The control chart shows a fairly good state of statistical control. The output of the process appears to be stable, consistent, predictable, and repeatable. The only thing that might trigger investigation is that there are eight averages in a row all above the centerline, beginning with sample 13 and ending with sample 20. It is not natural to see so many points in a row on the same side of the centerline. The probability of eight consecutive points on the same side of the centerline is $(\frac{1}{2})^8$, or 1/256, or a little less than 1 percent.

2. The process is aimed about 0.006 in higher than the target, which is 0.515 in. The process should be adjusted closer to the target.

---

*The A_2, D_4, and D_3 factors are in Table 5.1 at the end of this chapter.

The control limits are 3 standard deviations away from the centerline of the process for both the average and the range chart *for a process that is in statistical control.* Recall that 99.73 percent of the area under the normal distribution is between $+3\sigma$ and -3σ from the centerline. We should expect 99.73 percent of the sample statistics (averages and ranges in this example) to be within the calculated control limits when a process is in a *perfect* state of statistical control.

The distance from the centerline of the average chart of the control limits is 0.0046 in. This distance is found by multiplying the A_2 value (0.577) by the average range (0.008 in). That distance is mathematically equivalent to $3\sigma_{\bar{X}}$. The same calculation is performed for the lower control limit of the averages. Figure 5.4 graphically shows the distribution of individual measurements and the distribution of averages.

> QUESTION: Is there a piece of the puzzle missing? All this chart is telling me is that the planing of the board to a specified thickness is in control (stable, predictable, consistent, and repeatable) and that the thickness is running on the high side of the specification and should be adjusted downward, closer to the target of 0.515 in.
>
> ANSWER: There is a piece of the puzzle missing, and the name of the missing piece is the *capability* of the process, yet to be determined. The mea-

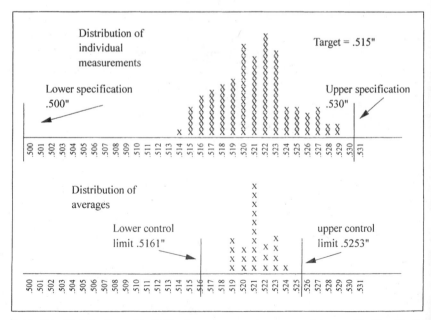

Figure 5.4 Distribution of individual measurements compared to specification limits and distribution of averages compared to control limits.

sures of *capability that will now be calculated are referred to as the C_p and the C_{pk} values for the process. First we calculate the C_p value. To determine the capability, we need to know the following values:

$$\bar{R} = 0.0080 \text{ in}$$

$$\text{Tolerance width} = 0.030 \text{ in}$$

$$d_2 \text{ for sample size of } 5 = 2.326$$

The first calculation will estimate the standard deviations of the individuals from the process that shows a good state of statistical control.

$$\hat{\sigma} = \frac{\bar{R}}{d_2} = \frac{0.0080}{2.326} = 0.0034 \text{ in} \tag{5.1}$$

$$C_p = \frac{\text{tolerance}}{6\hat{\sigma}} + \frac{0.030}{0.204} = 1.47 \tag{5.2}$$

Here $\hat{\sigma}$ denotes the estimated standard deviation of the individuals by using Eq. (5.1). The C_p value of 1.47 means that 147 percent of the width of the 6σ distribution of individuals will fit inside the specifications. The distribution of individuals will fit inside the specification limits with some room to spare.

The second aspect of evaluating the capability of the process involves the calculations for the C_{pk} value. The following values must be known in order to find C_{pk} for this process:

- Grand average $\bar{\bar{X}} = 0.5207$ in
- $\hat{\sigma} = 0.0034$ in
- Upper specification limit (USL) = 0.530 in
- Lower specification limit (LSL) = 0.500 in

$$Z_{\text{USL}} = \frac{\text{USL} - \bar{\bar{X}}}{\hat{\sigma}} = \frac{0.530 - 0.5207}{0.0034} = 2.74 \tag{5.3}$$

$$Z_{\text{LSL}} = \frac{\bar{\bar{X}} - \text{LSL}}{\hat{\sigma}} = \frac{0.5207 - 0.500}{0.0036} = 6.09 \tag{5.4}$$

$$Z_{\text{min}} = 2.74 \tag{5.5}$$

$$C_{pk} = \frac{Z_{\text{min}}}{3} = \frac{2.74}{3} = 0.91 \tag{5.6}$$

---

*Chapter 6 provides a more complete explanation of capability.

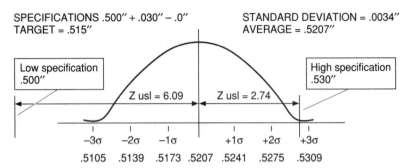

Figure 5.5 Graphical explanation of capability of board thickness.

Equations (5.3) and (5.4) are the Z formula that was explained in Chap. 3 except that the symbols are slightly modified. Equation (5.3) tells us that the upper specification limit is 2.74σ from the center of the process. Equation (5.4) tells us that the lower specification limit is 6.09σ from the center of the process. Recall from earlier discussion that the area under the normal distribution between +3σ and −3σ will have 99.73 percent of the measurements.

There is a small proportion of the product produced out of specification on the high end. Equation (5.6) gives us the C_{pk} value. The smaller of the two Z values is denoted by Z_{min}. In this case, 2.74 is the Z_{min} value, and that value is divided by 3. The final answer for C_{pk} is 0.91. Figure 5.5 graphically depicts the capability for this planing process.

Here is another way of explaining what C_{pk} is telling us: From the center of the process the nearest specification is 0.91 (91 percent) of 3 standard deviations. The process is set too close to the upper specification. There is much room for safety between the lower tail of the distribution and the lower specification. A very small portion of the tail of the high end of the distribution is outside the specification. According to the Z table, 0.31 percent of the product is outside the specification. It is not really the 0.31 percent out-of-specification product that we should be concerned with. What is worse is that there are not very many products close to the target, and conversely, many of the products are on the high side of the target of 0.515 in. This process is

1. In control
2. Running about 0.006 in high from the target of 0.515 in
3. Exhibiting the capability of consistently meeting specifications (C_p was 1.47)
4. Producing less than 1 percent of the product on the high end of the specification

Loss calculations

In past examples we determined the loss for the process as it exists now.

$$L = k[(\bar{y} - m)^2 + \sigma^2]$$

$\bar{y}$ = average m = target σ = standard deviation

$$L = \$1.00[(0.5207 - 0.5150)^2 + 0.0034^2]$$

$$= \$1.00(0.0057^2 + 0.000012) = \$0.0044$$

Notice that the great majority of the loss comes from the process running high off target, not due to the variation. The big savings (reduction in loss) is to be found not in preventing the 0.68 percent of product that is out of specification, but rather by targeting the process at 0.515 in. If the process were adjusted exactly to the target of 0.515 in with no reduction in variation, the C_p value would stay at 1.47 and C_{pk} would also be 1.47, as shown by the formulas below.

With the process not adjusted exactly at the target, the loss will be greatly reduced. The calculations for the loss are shown below:

$$Z_{USL} = \frac{USL - \bar{\bar{X}}}{\hat{\sigma}} = \frac{0.530 - 0.515}{0.0034} = 4.41$$

$$Z_{LSL} = \frac{\bar{\bar{X}} - LSL}{\hat{\sigma}} = \frac{0.515 - 0.500}{0.0034} = 4.41$$

$$Z_{min} = 4.41$$

$$C_{pk} = \frac{Z_{min}}{3} = \frac{4.41}{3} = 1.47$$

$$L = \$1.00[(0.5150 - 0.5150)^2 + 0.0034^2] = \$0.000012$$

The loss would be reduced by almost 75 percent just by centering the process on the target.

> QUESTION: It seems to me that a great deal of time, money, and effort was expended to determine that the process was running high. When I glanced at the individual measurements and the averages, it was obvious that the process should be adjusted down 0.005 or 0.006 in closer to the target. I think it was a waste of time to calculate the limits, plot all the averages and ranges, and calculate the capability.
>
> ANSWER: Occasionally I hear comments like this one. In some respects, your comment is appropriate. I agree that by reviewing the individual

measurements and scanning the averages you can estimate that the process is running high off target about 0.005 or 0.006 in, and making the adjustment closer to the target is the main thing that will improve the process. But your comment about wasting time, money, and effort to complete the evaluation of the process is wrong. First, not that much time, money, and effort is needed to complete the process evaluation. Second, much knowledge was gained by completing the evaluation. Until the control chart was completed, it was not known if the process was stable or unstable. The company did not know for sure if the process had the potential to meet the specifications. The C_p value of 1.47 tells us that the process can easily meet the specification requirements. Finally the C_{pk} value of 0.91 tells us that the third company goal has not been achieved. If a company considered itself world-class or a TQM company, it would not stop the evaluation of the process at the early stage, as you suggested.

QUESTION: One of the more common complaints from the manufacturing workers and supervisors is that these charts are just tightening up the tolerances, which are 0.030 in. Now this control chart is telling me to stay inside the control limits, and they are only 0.0072 in away from each other.

ANSWER: This idea used to be mentioned quite frequently in the past, but not so often in the last few years. Your question relates to the confusion between

1. Statistical control versus capability
2. Individual readings versus sample statistics ($\overline{X}R$, $\overline{X}\sigma$, $\widetilde{X}R$, XR_m)
3. Specification limits versus control limits

You commented about the control chart having limits that are only 0.0072 in away from each other, but the specification is 0.030 in wide. You have just confused control limits which are for sample statistics with specification limits which are for individual measurements. I think you were under the impression that the individual measurements should stay within the control limits which are 0.0092 in wide. That is not correct. The averages should stay within the control limits for a process to be in control.

5.6 Recalculation of Control Limits

Ask 10 people, When should control limits be recalculated? You will most likely hear two or three different answers. Before we list some typical responses, let's talk about some of the goals that we are trying to accomplish by using control charts for variables. Some of those goals are to

1. Improve the important quality characteristics so they are operating in a state of statistical control and are well centered on the target.

2. Improve the capability of quality characteristics so the goals for capability can be attained and high levels of customer satisfaction can be ensured.

With these goals firmly in mind, let us look at some typical responses to the question, When should new control limits be recalculated? You should recalculate new control limits whenever the current chart being filled out is full of current data. In other words, whenever the chart gets full, we recalculate the limits and put the new limits on the next chart to be filled out.

In a number of situations, this would be in disagreement with the goals mentioned earlier:

1. The process was on target but drifted off target. A shift in the centering of the process should trigger an out-of-control condition on the average or the individual chart. The recalculation of the control limits for a process that has drifted off target will silence the alarm (a point out of control).

2. If the variation in the process increased (worsened), causing the ranges to increase, then the next set of control limits for the new chart would be wider and these new limits would not accurately discriminate between common-cause and special-cause variation. This could potentially lead to *continual unimprovement*.

Another response heard today might be, "My software program recalculates the control limits every time a new sample is entered into the computer. Every time I enter in a new average $\bar{X}$ and range, the computer recalculates a new grand average and a new average range $\bar{R}$." These are referred to as *floating control limits*. This situation has many of the weaknesses mentioned above.

1. If the process drifts (averages drift), the computer accommodates this situation and the process could go off target but possibly remain inside the control limits which are just drifting with the process. (Of course, this assumes that the process was on target originally.)

2. If the variation in the process begins to increase, the ranges increase and the average range increases. Then all the control limits will become wider apart, and in this case, the chart will fail to properly identify special-cause variation when it occurs.

Another situation frequently arises when computers are used to calculate and plot control charts. An example is shown in Fig. 5.6. Rather than calculate the grand average $\bar{\bar{X}}$, the computer operator will force a line to be drawn at the target value (usually the nominal specification). The person doing this override has good intentions. The person thinks that if there is a line in the chart that is the target, then the process will be centered on that value. Usually that is not the case.

It is true that we would like the target and the grand average $\bar{\bar{X}}$ to be very close to the same value, but we must follow the accepted steps in calculating the limits.

112   Chapter Five

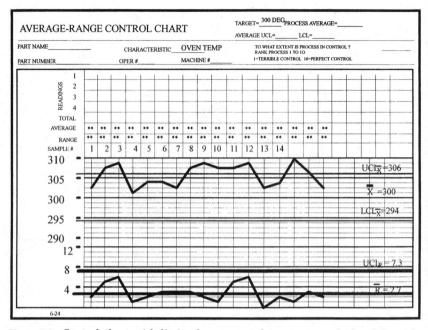

Figure 5.6 Control chart with limits drawn around target, not calculated from the process output.

If the target and the true central line on the average chart are very far from each other, this problem should be flagged by the machine operator, setup person, or supervisor. If it is not stressed at this time, that is very bad. This is an indication of not being involved in the process improvement effort. This problem will probably be scrutinized much later during the capability analysis because of an unsatisfactory C_{pk} value, which should get the proper person's attention. But manually drawing in a line on a control chart at the target value will do nothing to reduce cost or improve quality.

We have just discussed a few improper approaches to recalculating control limits. Now we discuss the proper method.

We should *recalculate control limits when the process has improved*. One may ask, How can I tell if the process has improved when I look at a control chart? There are two general ways to determine whether a process has improved:

1. If the averages get closer to the target
2. If the ranges decrease in value

If you take time to learn how to visually interpret control charts, before long you can develop the skill so that you can tell whether either of these improvements has occurred.

Shewhart Control Charts 113

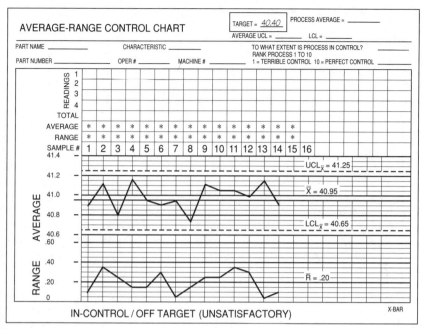

Figure 5.7 Control chart showing process in control, but off target a great deal (unsatisfactory situation).

Remember the objectives stated earlier:

1. Improve the process so that the process runs in a state of statistical control and is centered on the target.
2. Minimize common-cause variation so that there is good capability.

Example 5.1 Figure 5.7 is the initial chart for this process. The process was allowed to run the way it has been running in the past. We want to draw a baseline from where we started, so that we can see whether improvement has taken place. The control limits on this chart were calculated from the data on the same chart. The chart shows the process in a fairly good state of statistical control, but it is significantly off target.

This is not a satisfactory situation. In reality, we want the process to go outside *these* control limits. We want the process to be centered closer to the target. We should get together with whoever is in control of process adjustments and have the process adjusted closer to the target. Then we should recalculate new control limits to determine if the process is still in control after we have made the adjustment. If we were able to adjust closer to the target, the new grand average $\bar{X}$ would be closer to the target.

Once the process is on target and in control, we should freeze those limits unless we see a reduction in variation (ranges getting smaller). *We want to force an out-of-control condition to occur which will trigger investigation and will hopefully lead to a process improvement.*

5.7 Sampling Guidelines for Control Charts

There has been much discussion about which type of variable chart to use for a particular process and what the appropriate method for sampling the process should be to give the user meaningful information about process stability. There is no "cookbook" answer for all situations; it is not a black-or-white situation. The main areas addressed are (1) sample size, (2) sampling frequency, (3) number of measurements required, and (4) sampling for batch or continuous processes.

The following paragraphs should be considered as guidelines only—they are not cast in concrete.

1. The larger the sample size, the greater the cost to monitor the process. Some people will disagree with this statement. In some situations, if the operator has "idle time" in the machine cycle, there are no-cost penalties to the company if large samples are taken frequently. An attitude similar to this but more dangerous is the opposite: "The operator does not have time to fill out the chart, so we cannot use them" or "Only the inspector can fill out the chart."

2. The larger the sample size, the more sensitive the average* chart will be. The reason is that as the sample size increases, the less variation in the distribution of averages or medians there will be. This happens because as the sample size increases, the A_2 value used in the formulas decreases. Whatever the sample size, the control limits are $+3\sigma$ and -3σ from the center of the process.

3. If the distribution of individuals is not normally distributed, the sample size should be 4 or larger so that the distribution of averages will take the shape of a normal distribution. Do not think that this is improper manipulation of numbers. What we are talking about takes place naturally. It is a phenomenon known as the *central-limit theorem*. If this transformation did not happen, the power of the control chart would be negated. Figure 5.8 shows the central-limit theorem.

The central-limit theorem is a very powerful phenomenon. Glance at Fig. 5.8 and note two main things: (1) Despite the shape of the distribution of individuals, the distribution of averages will be the shape of the normal distribution if the sample size is at least 4. (2) The dis-

---

*This also applies to the distribution of medians, but not the distribution of individuals.

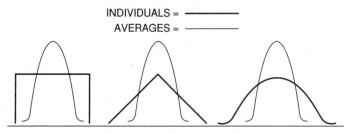

Figure 5.8 The power of the central-limit theorem.

tribution of averages in all the examples has reduced variation compared to the distribution of individuals. This means that the distribution of averages is more sensitive to change than the distribution of individuals is. We can notice if the process has changed more quickly if we compare distribution of averages to the control limits for that process. Some people might think we should evaluate individual measurements and compare the individual measurements to the specifications to determine if the process has changed. The goal is to prevent scrap, rework, customer complaints, and nonfunctional products. The strategy to accomplish this should be to bring the process into a state of control, center the process on the target, and then, if necessary, reduce common-cause variation.

Note: Some characteristics usually do not have the shape of a normal distribution, e.g., taper, flatness, machine efficiencies, concentricity, surface finish, end play, leak rate, and contamination level, to name just a few. Characteristics having specification limits that are expressed as a minimum or maximum will usually have a skewed distribution.

This does not mean that control charts cannot be used in the improvement of processes in which the distribution of individuals is not the shape of a normal distribution. When nonnormal distributions are charted, the goals should be to

1. Bring the process into statistical control.

2. Work toward aiming the process closer to the target value (usually zero, or infinitely high, depending on whether the characteristic has a minimum or maximum specification).

3. Improve the processes by reducing common-cause variation to the extent that C_p and C_{pk} are 1.67 minimum.

5.8 Sampling Schemes

There are many types of sampling schemes when you are using control charts. There is no true right or wrong method for sampling.

116   Chapter Five

Some processes naturally fluctuate differently from others. The goal is to find the best sampling scheme that will accurately identify special-cause variation so that we can efficiently work on improving a process or minimize common-cause variation.

Instant time sampling method

The instant time sampling method is the suggested sampling scheme. It implies taking a group of individual readings over a very short period. If possible, the readings should be consecutively produced. This sampling scheme will minimize the variation in the sample, but maximize the variation from sample to sample. This is shown in Fig. 5.9.

The roller-coaster contour and the sawtooth contour are common patterns of variation as time advances. If the sampling plan is three consecutive measurements, say, 1 min from each other, probably the measurement will show little variation within the sample. This sampling scheme will show greater variation from sample average to sample average. These smaller ranges will cause the average range to be smaller and the control limits to be closer to each other on both charts.

Over-time sampling

The over-time sampling (Fig. 5.10) suggests that the individual measurements should be taken with some time between readings (possibly having 5 to 30 min between readings). This is not improper manipulation of statistics. Some processes will display very little variation over a short period (5 to 30 min). In turn, the ranges will be quite small, and if

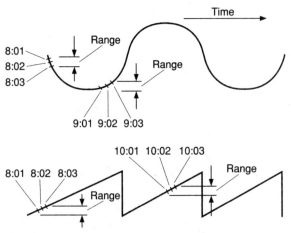

Figure 5.9 Instant time sampling scheme.

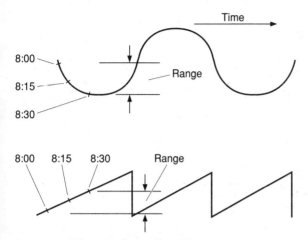

Figure 5.10 "Over-time" sampling scheme.

we recall the mathematics involved in the control limit calculations, the average range $\overline{R}$ affects the control limits on both charts. For practical purposes, the control limits might be unrealistically tight. Using the "over a period of time" method of sampling, one would expect the ranges to be somewhat larger, which will cause the control limits to be farther away from one another. The averages will tend to exhibit less variation when over-time sampling is used.

Figure 5.11 is an example of a control chart where the use of the instant time sampling method causes problems. A company that has a stamping operation has decided to use a control chart to measure the distance from the part's edge to the centerline of a hole. The specification is 2.500 ± 0.010 in. The total tolerance is 0.020 in. The target is 2.500 in. There was variable gaging at the press. The sampling plan calls for three consecutive parts to be measured every 30 min. The measurements on the control chart are coded from a reference point of 0 = 2.500 in. The measurements are in 0.001-in increments (for example, +1.5 = 2.5015 in).

Notice that the ranges in the sample are quite small. The calculated control limits are shown here:

$$\overline{\overline{X}} = \frac{\Sigma \overline{X}}{k} = \frac{50.83}{25} = 2.03 = 2.50203 \text{ in}$$

$$\overline{R} = \frac{\Sigma R}{k} = \frac{29}{25} = 1.16 = 0.00116 \text{ in}$$

$$\text{UCL}_{\overline{X}} = \overline{\overline{X}} + A_2 \overline{R} = 2.50203 + 1.023(0.00116) = 2.5032 \text{ in}$$

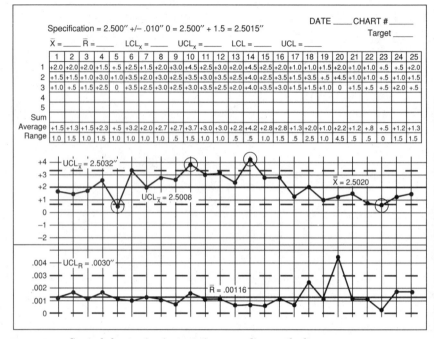

Figure 5.11 Control chart using instant time sampling method.

$$\text{LCL}_{\bar{X}} = \bar{\bar{X}} - A_2 \bar{R} = 2.50203 - 1.023(0.00116) = 2.5008 \text{ in}$$

$$\text{UCL}_R = D_4 \bar{R} = 2.574(0.00116) = 0.0030 \text{ in}$$

$$\text{LCL}_R = D_3 \bar{R} = 0(0.00116) = 0$$

The range chart shows one point out of control, sample range number 20. The averages are wildly out of control based upon this sampling scheme. There are four averages outside the control limit samples: 5, 10, 14, and 23. Also there are many points out close to the control limits but few averages close to the centerline of the average chart. At first glance, we consider this process to be a category 3 process (out of control but consistently meeting specification). It appears that there is a great deal of special-cause variation influencing the process.

This is a prime example of falsely thinking that the process is out of control when it really is not. This situation is called a *type I statistical error*. Sometimes this error is called the *producer's risk*. The producer thinks the process is out of control when really it is not. The sampling scheme is not appropriate for the variation in this process. If this chart were put at the press and the operator and support departments were assigned to keep the process in control, great problems

Shewhart Control Charts 119

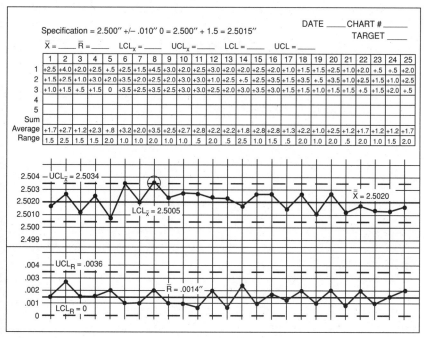

Figure 5.12 Control chart using over-time sampling scheme.

and much frustration would ensue. This would have a negative effect upon the process improvement effort. Piece-to-piece variation within the sample is quite small. Time-to-time variation is larger (from sample average to sample average).

Figure 5.12 is an average and range chart showing the same process 2 days later with a different sampling scheme. The sampling scheme now is to record one reading every 10 min and then calculate and plot averages and ranges with a sample size of 3. An average and range were plotted every 30 min. This chart shows a much better state of statistical control than the first chart and does a better job of identifying special-cause variation correctly.

$$\overline{\overline{X}} = \frac{\sum \overline{X}}{k} = \frac{51.2}{25} = 2.05 = 2.5020 = 2.5020 \text{ in}$$

$$\overline{R} = \frac{\sum R}{k} = \frac{35.5}{25} = 1.4 = 0.0014 \text{ in}$$

$$\text{UCL}_{\overline{X}} = \overline{\overline{X}} + A_2 \overline{R} = 2.5020 + 1.023(0.0014) = 2.5034 \text{ in}$$

$$\text{LCL}_{\overline{X}} = \overline{\overline{X}} - A_2 \overline{R} = 2.5020 - 1.023(0.0014) = 2.5005 \text{ in}$$

$$\text{UCL}_R = D_4 \overline{R} = 2.574(0.0014) = 0.0036 \text{ in}$$

$$\text{LCL}_R = D_3 \overline{R} = 0(0.0014) = 0$$

Shown below is the comparison of the calculated statistics and control limits for the two sampling schemes.

Statistic	Instant sampling (in)	Over time (in)
$\overline{\overline{X}}$	2.5020	2.5020
$\overline{R}$	0.00116	0.0014
$\text{UCL}_{\overline{X}}$	2.5032	2.5034
$\text{LCL}_{\overline{X}}$	2.5008	2.5005
UCL_R	0.0030	0.0036

From a statistical control point of view, neither of the control charts is in a perfect state of statistical control. The instant time sampling chart does not exhibit the expected pattern of many averages (68 percent) falling close to the centerline. Also, there are four averages outside the calculated control limits. The range chart does show one point outside the upper control limit for the ranges.

The over-time chart shows some significant differences just based upon changing the sampling scheme:

1. The average range is about 20 percent larger than the instant time sampling chart.

2. The upper control limit for the ranges is greater by 20 percent.

3. The distances that the average control limits are from one another is 20 percent greater than with instant time sampling.

4. The averages are not swinging as wildly because the over-time sampling smooths the averages to some extent.

> QUESTION: Is this last chart not still showing a process that is out of control? The first goal of statistical control continuously mentioned has not yet been attained. I think there would still be much frustration from factory workers trying to keep the process in control. Also, I think that this is a way to manipulate the type of control chart so that the chart looks good (in control).
>
> ANSWER: Your first statement is correct: The process is not in perfect control, and it will be a challenge to improve this process so it is in perfect statistical control in this situation. This may not be a great opportunity for improvement in quality, costs, and customer satisfaction. The tolerance for this characteristic is 0.020 in. Remember that any measurement between $+10$ and -10 is in specification. Just by visually scanning the individual measurements on either control chart, we see that the process is centered quite well and very easily meets the specifications. Figure 5.13 shows the distribution of individuals for both instant time and over-time sampling.

Shewhart Control Charts 121

```
Over a period of time sampling
distribution of individual readings
Specification 2.500"  +/- .010"
readings are coded from zero:

+4.5 | X
+4.0 | X
+3.5 | XXXXXXX
+3.0 | XXXXXXXXX
+2.5 | XXXXXXXXXXXXXXX
+2.0 | XXXXXXXXXXXX
+1.5 | XXXXXXXXXXXXX
+1.0 | XXXXXXXX
+ .5 | XXXXXXXX
   0 | X——————————— 0 = 2.500"
```

```
Instant time sampling
distribution of individual readings
Specification 2.500"  +/- .010"
readings are coded from zero:

+4.5 | XX
+4.0 | XX
+3.5 | XXXXXXXX
+3.0 | XXXXXXXX
+2.5 | XXXXXXXXXX
+2.0 | XXXXXXXXXXX
+1.5 | XXXXXXXXXXXX
+1.0 | XXXXXXXXXXX
+ .5 | XXXXXXXXX
   0 | XX—————————— 0 = 2.500"
```

Figure 5.13 Histogram showing individual measurements for both instant time chart and over-time chart.

There are still many alternatives about how to monitor the process and not chase problems that do not really exist (type I error). These are some alternative techniques:

  Use an individual and moving range control chart.
  Use a moving average and moving range control chart.
  Use a control chart with modified limits.
  Use a tool other than a control chart, such as a run chart, histogram, precontrol chart, or check sheet.

The approach should be to monitor the process by using a technique that is not too restrictive, but one that will attract attention before trouble is encountered.

Regarding the second comment about manipulating the chart to make it look in control, there is always the chance that making the chart look good will be the objective of some individuals who fail to understand process improvement. Some people will take the hard-line approach about what charting technique and sampling scheme to use. As a person knows more about the broad choice of tools and techniques available, that person will show flexibility in managing the process improvement effort. The charting techniques used should ensure high levels of quality but not cause interruptions of a manufacturing process for no good reason. It is unlikely that this example of charting this quality characteristic is very high on the list of opportunities for improvement within this company. We should really stop this discussion about the sampling technique or the type of chart used and reconsider our priorities for improvement.

Caution! Do not interpret the preceding discussion to mean that the sampling scheme should *always* be the over-time method. The suggested method of sampling is the instant time sampling method, if practical.

Number of measurements

The final decision about the number of measurements should be made only after some of the following factors have been taken into consideration:

- How big is the opportunity for improvement?
- What is the production rate?
- What is the cost of making this measurement?
- How long does it take to make this check?
- What is the changeover rate from product type to product type?

Mil-Std 414

There are suggested guidelines as to the sample size and frequency for a particular lot size. Most of the work developing these guidelines was done in the context of traditional manufacturing situations where discrete individual products are produced, which is not the case in batch process or continuous process operations. Remember that these are *suggested* guidelines only.

Example 5.2 The final packaging of cans of copper concentrate has two different lines. Each line produces 1800 cans per 8-h shift. (The lot size is considered to be 1800 units.)

Normal Inspection, Level 4

Lot size	Sample size
66–110	10
111–180	15
181–300	25
301–500	30
501–800	35
801–1300	40
1301–3200	50
3201–8000	60
8001–22,000	85

The table on the facing page tells us that we should have 50 measurements per shift. This is only a guideline. The factors mentioned earlier should be weighed and then a revision made if it is warranted. Obtain the approximate number of measurements in any one of many combinations:

- 50 samples of 1 during the shift
- 25 samples of 2 during the shift
- 16 samples of 3 during the shift
- 12 samples of 4 during the shift
- 10 samples of 5 during the shift
- 8 samples of 6 during the shift

There are pros and cons for each of the above options. Suppose that an average and range chart will be used. Remember that the larger the sample size on the control chart, the more sensitive the average chart will be. But the responsiveness (time lag to plot a point) worsens as the sample size increases. The two considerations of sensitivity and responsiveness are in conflict with each other. If this conflict between sensitivity and responsiveness causes problems, there are a few ways around it. An individual and moving range chart is responsive but not sensitive. The average and range chart with large samples is more sensitive but not very responsive. The best of both worlds may very well be the moving average and moving range chart. This chart will be explained later in this chapter.

5.9 Sampling for Continuous Processes or Batch Processes

Suppose that a petroleum product is being produced through a production process at a constant rate of 14 gal/min. One blending tank has a volume of 350 gal. The residence time for the product at this stage is 25 min. Residence time must be a consideration when a sampling system for a continuous process is developed. The product is processed continuously, and experience has proved for a process similar to this it is good to take an individual measurement about once every 25 or 30 min or more.

The traditional method for calculating control limits is to multiply the within-subgroup variation (average range $\bar{R}$) by the Shewhart factor to determine the subgroup-to-subgroup variation. Occasionally, this method is not appropriate to determine the control limits. Recall that the purpose of the control limits is to identify any special-cause variation.

Two types of statistical errors can occur:

Type I error: This type of error will falsely trigger an out-of-control condition. This really is not an out-of-control condition. This type of error is referred to as the *producer's risk*.

Type II error: This type of error occurs when special-cause variation is present in the process but the control chart for some reason does not identify the situation properly.

Sometimes when we are monitoring continuous processes, using the traditional method for calculating control limits, a type I error occurs. We think the process has gone out of control when it really has not.

The traditional method for calculating the limits for the average chart is

$$\text{UCL}_{\overline{X}} = \overline{X} + A_2 \overline{R}$$

It is both costly and frustrating trying to find the cause of an out-of-control condition when there is none. This problem can be prevented by using another method for calculating the control limits which employs the variation from subgroup to subgroup to determine the control limits for the chart that measured central tendency (average, median, moving average, or individual). Remember, these control limits are $+3\sigma$ and -3σ from the centerline.

A nontraditional method for calculating control limits uses

$$\text{UCL}_{\overline{X}} = \overline{X} + 2.33\sigma_{\overline{X}}$$

Two main differences arise from calculating the control limits by this nontraditional method:

1. We are using variation from subgroup averages to subgroup averages to determine the control limits, rather than using the average variation within the subgroup.
2. The control limits in the nontraditional method are $+2.33\sigma$ and -2.33σ from the centerline rather than the standard $+3\sigma$ and -3σ. Many quality professionals say that this approach to calculating control limits for continuous processes works well.

Note: The control limits for the range chart are calculated in the traditional way.

5.10 Individual and Moving Range Chart

Since the development of control charts by Dr. Walter Shewhart in the 1920s, many other types of variable control charts have been developed for certain situations. A control chart for individuals and moving ranges is shown in Fig. 5.14.

In certain processes, it may not be economically practical or logical to monitor a process by evaluating sample averages and ranges. One of many options would be to monitor individual measurements and moving ranges. This situation is appropriate when the testing is very

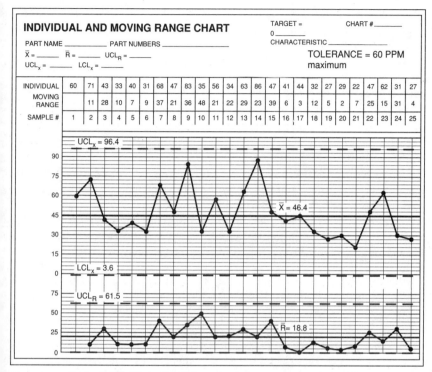

Figure 5.14 Control chart for individuals and moving ranges.

expensive or possibly for destructive testing, low production rates, or in a continuous process where the output is homogeneous.

There are some inherent weaknesses in this type of control chart:

1. Charts for individuals do not give us as much "sensitivity" or early warning as charts for averages. The process can drift more before an out-of-control condition is triggered with an individual chart than with an average control chart. The reason is that the control limits for individual charts are 3σ of the individuals from the centerline of the chart whereas the control limits for the average chart are 3σ of the averages from the centerline of the chart. Remember that the distribution for averages will have less variation than the distribution for individuals for the same process.

This is not improper manipulation of numbers; the control limits for the individuals chart should be further apart than the control limits for averages because the distribution of individuals has greater variation than the distribution of averages.

2. Caution must be exercised in the visual pattern analysis of a chart for individuals if the distribution of individuals is not the general shape of a normal distribution. Most of the rules for visual pattern

analysis are based upon the assumption that the distribution being analyzed is the general shape of the normal distribution. If the distribution of individuals is badly skewed or is any other nonnormal shape, then there is a chance that the control chart is not properly flagging special-cause variation. If the distribution of individuals varies greatly from the shape of the normal distribution, then the rules for visual pattern analysis should not be enforced so restrictively and the type of control chart or sampling scheme should be modified. This problem can usually be eliminated by using a moving average and moving range chart.

3. Charts for individuals do not isolate piece-to-piece repeatability of the process. Since there is only one item per subgroup, it might be wiser to use some type of average and range chart.

4. There should be at least 50 measurements before control limits are calculated and before the capability of the process is estimated. This is so because there is only one reading per sample.

Example 5.3 A cosmetic manufacturer is measuring the contaminant level in a product. The specification is 60 ppm maximum. Measurements are taken every 15 min and are recorded. The past procedure requires that if the measurement is beyond the specification limit, the batch of product produced during that 15 min is reprocessed through the sterilizing operation. The current culture within the company has stressed production over quality. It was believed that the penalty for reprocessing due to high contaminants was small compared to the penalty of reducing the throughput rate. The focus was not to go over 60 ppm, rather than to keep driving the contaminant rate down closer to the target of zero. In this situation, the goal is to drive the process out of control on the low side, since this is a smaller-is-better (SIB) type of characteristic, but not reduce production rates to achieve higher purity. The calculations for the control limits for the individual and moving range chart are as follows:

$$\bar{X} = \frac{\Sigma X}{k} = \frac{1159}{25} = 46.4 \text{ ppm}$$

$$\bar{R} = \frac{\Sigma R}{k-1} = \frac{451}{24} = 18.8 \text{ ppm}$$

$$\text{UCL}_X = \bar{X} + E_2 \bar{R} = 46.4 + 2.660(18.8) = 96.4 \text{ ppm}$$

$$\text{LCL}_X = \bar{X} - E_2 \bar{R} = 46.4 - 2.660(18.8) = -3.6 \text{ ppm}$$

$$\text{UCL}_R = D_4 \bar{R} = 3.267(18.8) = 61.5$$

$$\text{LCL}_R = D_3 \bar{R} = 0(18.8) = 0$$

None of the 25 individual measurements or the 24 moving ranges are outside the calculated control limits. There are some unnatural patterns—there is a run of nine ranges above the average range. The individual chart shows an unbalanced pattern. Notice that the individuals below the centerline appear bunched close to the centerline, whereas the individuals above the centerline are rather stretched out. Recall from earlier in the book that control charts for individuals should not be used if the distribution of individuals differs greatly from the normal distribution. The rules for visual pattern analysis do not really apply as strongly. There is the likelihood that either special-cause variation will not cause an out-of-control condition (consumer's risk) or that common-cause variation will trigger an out-of-control condition (producer's risk). If the workers have learned how to perform visual pattern analysis of the control charts, there will be some confusion and frustration due to this situation.

There are a few things unique to the individual control chart that should be explained:

1. The central line on the individual chart is the average $\overline{X}$, not the grand average $\overline{\overline{X}}$.
2. Even though one measurement is recorded and plotted at a time, we consider the sample size to be 2 in this example, because the moving ranges are calculated by comparing two individual measurements.
3. The factor used to calculate the limits for the individual chart is not A_2, which was used for the control chart for averages. A factor E_2 is used in the calculations for individuals. In this example the sample size is 2, and the E_2 factor is 2.660. The factors used in the calculations for the individuals chart are found in Table 5.1 at the end of the chapter.
4. There is one less moving range than there is individual measurement on the chart. In the calculation for the average of the process, the sum of the individuals is divided by 25, whereas the average range is determined by dividing the sum of the ranges by 24.

Notice that the lower control limit for individuals is a negative number (-3.6 ppm). It is impossible to have a negative contamination level. If there were no mistakes in the plotting of the data or the calculations, it is a strong indication that the distribution of individuals is skewed, not a normal distribution.

Figure 5.15 shows the histograms of individuals compared to the specification limits and the distribution of individuals compared to the control limits. *Note:* In many previous examples this same type of graphic was shown, except that the type of chart being used was an average chart, not an individual chart. Also recall that the distribution of averages shows less variation than the distribution of individuals does.

Capability calculations

The standard method for determining capability should not be followed at this time for two reasons:

1. The individual and moving range chart cannot accurately determine whether this process is in a state of statistical control because the distribution of individual measurements is a skewed distribution.

2. The standard method for determining capability would be misleading since the distribution of individuals is not the shape of the normal distribution. Capability should be determined by using the normal probability paper (NPP) method explained in Chap. 3.

Example summary

The chart for individuals should be given serious consideration when the process does not lend itself to consecutive-type sampling. The in-

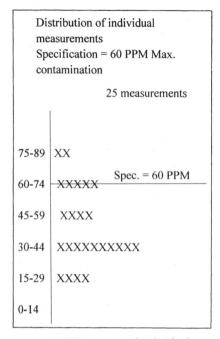
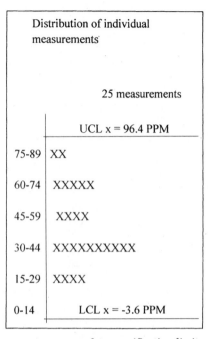

Figure 5.15 Histograms of individual measurements compared to specification limits and distribution of individuals compared to control limits (contaminant level, maximum specification).

dividual chart is not as sensitive as an average chart or a median chart. The individual chart should not be used if the distribution of individual measurements differs greatly from the general shape of the normal distribution. The moving average and moving range chart would be more appropriate for this type of process.

5.11 Moving Average and Moving Range Chart

The previous discussion focused on individual and moving range charts. The major advantage of this type of control chart is that it accommodates processes that do not lend themselves to the use of average and range charts or of median and range charts. Batch processes and continuous processes often fluctuate in such a fashion that the standard sampling techniques (consecutive) will have control limits so close to each other that the limits are just not practical. Life will be miserable for the manufacturing and support departments. The operator will have great difficulty keeping the averages within the calculated control limits.

When calculating the limits for moving averages, we use the standard A_2 factor rather than E_2. Some advantages of using the moving average and moving range chart are that

1. The control limits for the moving averages chart will be closer to each other, which increases the sensitivity over that experienced with individual charts.
2. The moving average chart can be used to chart any shape of distribution of individuals since the sample size is 4 or greater (central-limit theorem).

Figure 5.16 is the moving average and moving range control chart with the same measurements as were recorded on the individual and moving range chart.

Since the distribution of individuals is not the shape of the normal distribution, the sample size will be 4. This is done to ensure that the distribution of individuals more closely follows the shape of the normal distribution. Setting the sample size to 4 will more accurately identify special-cause variation or any other unnatural pattern of variation.

$$\bar{\bar{X}} = \frac{\sum \bar{X}}{k} = \frac{1019.9}{22} = 46.4 \text{ ppm}$$

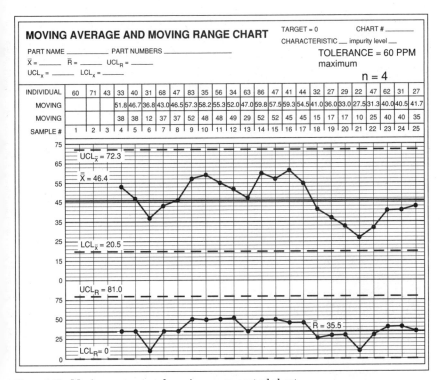

Figure 5.16 Moving average and moving range control chart.

$$\overline{R} = \frac{\Sigma R}{k} = \frac{781}{22} = 35.5 \text{ ppm}$$

$$\text{UCL}_{\overline{X}} = \overline{\overline{X}} + A_2 \overline{R} = 46.4 + 0.729(35.5) = 72.3 \text{ ppm}$$

$$\text{LCL}_{\overline{X}} = \overline{\overline{X}} - A_2 \overline{R} = 46.4 - 0.729(35.5) = 20.52 \text{ ppm}$$

$$\text{UCL}_R = D_4 \overline{R} = 2.282(35.5) = 81.0 \text{ ppm}$$

$$\text{LCL}_R = D_3 \overline{R} = 0(35.5) = 0$$

Comments about the calculations

Even though there are 25 individual measurements on the control chart, there are only 22 moving averages and moving ranges. With the sample size set at 4, there was no average or range calculated for the first three measurements.

The centerline is the grand average $\overline{\overline{X}}$, not the average $\overline{X}$. Factor A_2 is used in the calculations for the moving average chart, not E_2, which was used in the calculations for the individual chart. This will cause the control limits for the moving averages to be tighter, which will give us more sensitivity than we would have if we were using a chart for individuals and moving ranges. Some people will feel as though all we are doing is manipulating the data-gathering method to make the chart look the way we want it to. But this is not really the case. The distribution of individual measurements of the product will not change if we change the type of control chart used. The scrap and/or rework rate will not increase or decrease. The final customer will not see a difference in the level of the quality of the product unless there is improvement in the process. It is true that there will be some differences when the control limits for individuals are compared with the control limits for the moving averages.

Statistic	$X - R_m$ (ppm)	$\overline{X}_m - R_m$ (ppm)
Average	46.4	46.4
$\overline{R}$	18.8	35.5
UCL	96.4	72.3
LCL	−3.6	20.5
UCL_R	69.0	81.0

Both charts have the same average. The average range for the moving average and moving range chart is much greater than that for the individual and moving range chart. This is to be expected because the sample size is 4 in the moving average and moving range chart versus 2 for the individual and moving range chart. The control limits for the moving averages are closer to each other than the control limits for the individuals are. Both sets of control limits are ±3σ from the centerline. The upper control limit for the range chart is greater for the moving average and moving range chart (sample size = 4) than for the individual and moving range chart (sample size = 2). This is proper because the ranges will be greater with a sample size of 4 than with a sample size of 2.

Admittedly, the look of the control chart will vary somewhat depending on the way we sample and the type of control chart we use. The main purpose of the control chart is to accurately determine whether only common-cause variation is affecting the process or some special-cause variation is affecting the process. The moving average and moving range chart does a better job of accomplishing this goal.

All 22 moving averages and 22 moving ranges are within the control limits. This is a sign that leads us to believe that the process is in a fairly good state of statistical control. When we do visual pattern analysis of the moving average chart, we notice that fewer than expected of the moving averages are close to the centerline. There are more moving averages out toward the control limits than expected. The moving average chart seems to show a cyclic pattern, which is not natural; notice that the last eight moving averages are all below the centerline—this run is unnatural. Recall that this quality characteristic—contamination level—is a smaller-is-better characteristic. The run of the last eight averages should be investigated to find out what caused this *good* unnatural situation. Possibly this run of the last eight averages could be a result of a low product throughput rate. We should also be able to track the production rates during the times of the contamination levels being charted. A scatter plot would be a good technique to use to ascertain if there is a relationship between the production rate and the contamination levels.

The estimate of capability will differ according to which charting method is used. But the important point to understand is that there is not one statistical tool that should be used for all applications. Many times, traditional control charts are not the appropriate tools to use for all processes.

5.12 Modified-Limits Chart

A modified-limits control chart should be considered when the following situation is encountered: The product produced by this process is very, very seldom out of specification, and the level of customer satisfaction is high. The customer who uses this product requires some type of control chart to be employed to monitor the process. An average and range chart was used in the past, but there was no success in finding the cause of out-of-control conditions and preventing their recurrence. When the process was out of control, there was never any out-of-specification product produced. The customer wants some type of statistical tool to be used to monitor the process, but none of the parties feel as though there is a need for process improvement. It might be worthwhile to ask the customer if a histogram, a run chart, or a simple check sheet could be used to monitor the process. This situation sounds very similar to the stamping example earlier in the chapter. In that example an average and range chart was used to monitor the process. The sampling scheme was to measure three consecutively produced products. The average control limits were so close to each other that they were not practical.

An alternative sampling scheme improved this situation: Record three measurements with a period of time between the measurements to allow the process to show some variation. The process was closer to a state of statistical control when over-time sampling was used, and the control limits seemed more practical.

An example of a control chart with modified limits is shown in Fig. 5.17 with the same measurements as in the stamping example (the instant time sampling scheme).

The term *control limit* is not used for the average chart when modified limits are used; the modified limits are referred to as upper and lower *rejection limits* for the averages ($URL_{\bar{X}}$, $LRL_{\bar{X}}$).

Notice that the grand average calculation is not performed for this type of control chart. A modified-limits control chart does not have a centerline. The objective is to keep the sample averages inside these rejection limits. We are not as concerned about unnatural patterns of variation on the control chart.

Factor A_2 is not used in the calculations for the rejection limits for the averages. The standard deviations of the individuals must be calculated by using some type of computer program, a handheld calculator, or the cumbersome longhand method explained in Chap. 3. Here the standard deviation of the 75 individual measurements is 1.2 (0.0012 in). The averages will be charted, so we must determine a standard deviation for the averages. The central-limit theorem states

$$\sigma_{\bar{X}} = \frac{\sigma_X}{\sqrt{n}} = \frac{0.00116}{\sqrt{3}} = 0.0007 \text{ in}$$

Thus $3\sigma_{\bar{X}}$ is 0.0021 in. The specification for this quality characteristic is 2.500 ± 0.010 in.

$$URL_{\bar{X}} = USL - 3\sigma_{\bar{X}} = 2.510 - 0.0021 = 2.5079$$

$$LRL_{\bar{X}} = LSL + 3\sigma_{\bar{X}} = 2.490 + 0.0021 = 2.4921$$

$$UCL_R = D_4 \bar{R} = 2.574(0.00116) = 0.0030$$

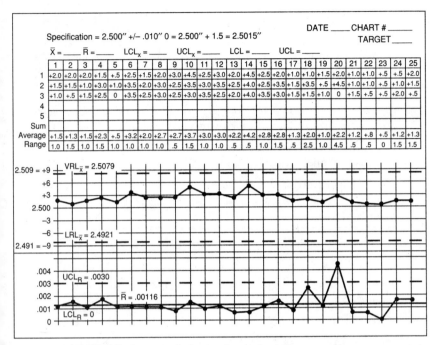

Figure 5.17 Modified-limits control chart using instant time sampling measurements from stamping operation (see Fig. 5.11).

$$\text{LCL}_R = D_3 \overline{R} = 0(0.00116) = 0$$

QUESTION: Are these rejection limits not very close to the specification limits? These rejection limits for the averages will most certainly reduce the problem of looking for out-of-control conditions; then there is no real reason to improve the process. But now I have a different concern. I am concerned that if the instructions for the production people are to keep the averages within these warning limits, then out-of-specification product could be produced even with the average within the rejection limits. In other words, out-of-specification product is produced, but the alarm does not go off. Do you have any suggestions?

ANSWER: There is much flexibility in the use of this type of control chart. The rejection limits for the averages that we just calculated were 3σ of the averages from the specifications. By visually analyzing the modified-limits chart, it appears that the warning limits could easily be 6σ in from the specifications and not cause unnecessary interruptions to the process. In this case the warning limits would be

$$\text{URL}_{\overline{X}} = 2.5058 \text{ in} \qquad \text{LRL}_{\overline{X}} = 2.4942 \text{ in}$$

These warning limits will increase the likelihood that if an average were outside these limits, no out-of-specification product would be produced.

The major differences in the two types of charts are as follows:

1. The modified limits for the averages (rejection limits) will be significantly farther from each other than the standard control limits would be. (This is for the average chart only; there is no difference in the range chart.)
2. The modified-limits chart does not have a centerline, so there is not the pressure to keep the averages aimed toward the centerline.
3. The modified-limits chart cannot actually discriminate between common- and special-cause variation.
4. The standard method for calculating capability cannot be used because the formulas were based upon the process being in true statistical control.

The modified-limits chart should be considered when the following occurs:

1. The process can easily meet the specification limits, and customer satisfaction levels are high.
2. It is very difficult or not economically practical to keep the process averages in true statistical control.
3. A decision has been made to use some type of control chart rather than a histogram, a check sheet, or one of the other tools of statistical problem solving.
4. The range chart shows a good state of statistical control.

5.13 Multistream Control Chart

This variation of the standard control chart should be considered when there are multiple streams of the product even though there is only one machine or processing line. This situation arises when there are multiple spindles, fixtures, cavities, dies, etc. At the start of evaluating a process with multiple streams, we must be able to determine whether all the streams are aimed at the same target and exhibit uniform variation from stream to stream (Fig. 5.18).

The multistream control chart is set up to monitor multiple streams. When just one control chart is used, you will notice that where we enter the individual readings, we have made the notation *fixture number*. It is extremely important to explain to the people gathering the readings to record the reading from fixture 1 in the space for fixture 1, and so on. In this example, process improvement may require the retargeting of one fixture or effort to reduce variation in one of the fixtures. The measurements must be traceable to the process stream from which they came.

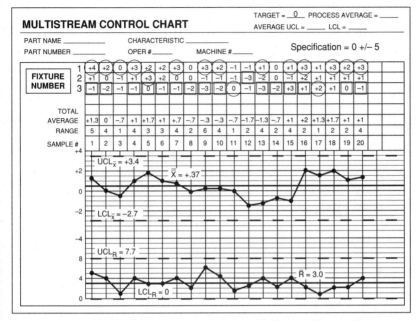

Figure 5.18 Control chart for multistream process with high and low readings marked.

The calculations for the control limits are as shown:

$$\bar{\bar{X}} = \frac{\Sigma \bar{X}}{k} = \frac{7.3}{20} = 0.37$$

$$\bar{R} = \frac{\Sigma R}{k} = \frac{60}{20} = 3.0$$

$$\text{UCL}_{\bar{X}} = \bar{\bar{X}} + A_2 \bar{R} = 0.37 + 1.023(3.0) = 3.4$$

$$\text{LCL}_{\bar{X}} = \bar{\bar{X}} - A_2 \bar{R} = 0.37 - 1.023(3.0) = -2.7$$

$$\text{UCL}_R = D_4 \bar{R} = 2.574(3.0) = 7.7$$

$$\text{LCL}_R = D_3 \bar{R} = 0(3.0) = 0$$

The range chart shows a good state of control. The average chart has no points outside the control limits. However, there happen to be two unnatural patterns: (1) There is a run of eight consecutive points below the centerline. (2) There is an absence of any points close to the control limits for the average chart. There are no points close to the limits because fixture 1 is adjusted high and fixture 3 is adjusted low. The average and the range of each fixture are shown on the next page.

Fixture no.	Average	Range
1	+1.65	5
2	+0.10	6
3	−0.70	5

Because the fixtures are set at different levels, the ranges will be large, the average range will be large, and the control limits for the averages will be incorrectly far apart. The averages will "hug" the centerline because the effects of fixture 1 set high and fixture 3 set low cancel on the average chart. If the person analyzing this control chart did not notice the unnatural pattern and its cause, then the calculated capability would be somewhat misleading, on the pessimistic side. Shown here are the calculations for the capability:

$$\hat{\sigma} = \frac{\overline{R}}{d_2} = \frac{3.0}{1.693} = 1.77$$

$$C_p = \frac{\text{tolerance}}{6\hat{\sigma}} = \frac{10}{10.63} = 0.94$$

$$Z_{\text{USL}} = \frac{\text{USL} - \overline{\overline{X}}}{\hat{\sigma}} = \frac{5 - 0.37}{1.77} = 2.66$$

$$Z_{\text{LSL}} = \frac{\overline{\overline{X}} - \text{LSL}}{\hat{\sigma}} = \frac{0.37 - (-5)}{1.77} = 3.03$$

$$C_{pk} = \frac{Z_{\min}}{3} = \frac{2.66}{3} = 0.89$$

The C_p value of 0.94 leads us to believe that the 6σ distribution of measurements will not quite fit between the specification limits of +5 and −5. The theoretical proportion of product out of specification on the high side is 0.39 percent. The theoretical proportion of product out of specification on the low side is 0.12 percent. The highest reading observed is +4 on samples 1 and 5; and the lowest reading observed is −3, in samples 9, 13, and 15. This seems to be a minor conflict between the theoretical and observed. The main point being made is that fixture 1 should be adjusted down and fixture 3 adjusted up closer to the target; then the variation in the product from this three-stream process will be reduced.

> QUESTION: It seems that somehow the fact that fixture 1 is set high and fixture 3 is set low should have been discovered earlier without all this effort. Do you have any suggestions as to how to accomplish this before calculating the limits and estimating capability?

ANSWER: Yes, there are numerous ways of doing this. One option is to set up a run chart for each of the fixtures separately. Then evaluate the run charts for targeting and variation. Another option is to use one control chart but analyze the measurements differently. In the past we have evaluated the readings in the vertical direction, but in this situation we want to be able to evaluate the readings in the horizontal direction. We want to be able to determine whether the different fixtures are aimed differently or repeat differently. This can be done in different ways, depending upon the accuracy needed. This can be done by evaluating the first sample and drawing a *circle* around the highest reading. Repeat this for all the samples. It is obvious that fixture 1 is running high. Now go back to the first sample, find the lowest reading in that sample, and draw a *box* around the lowest reading. Repeat this for all the samples shown in Fig. 5.18.

It is obvious that fixture 3 is running low and fixture 1 is running high. There is actually no need to calculate the control limits for this process. Simply adjust the two extreme fixtures (1 and 3) closer to the target of zero. Once that has happened, the distribution of product will be more uniform, and it follows that the ranges will be smaller and the control limits will be closer to each other. Once the two fixtures are moved closer to the target, it makes sense to calculate control limits for the process.

Some people feel that if there are multiple streams in one machine, there should be one control chart for each stream. This approach can snowball into an extremely large number of control charts, and if we are not careful, we will spend all day doing the calculations for all the charts rather than evaluating the charts and *solving problems.*

5.14 Control Charts for Low-Volume Applications

More and more people are finding out that it makes more sense to chart the "process" than the product in the world of smaller and smaller lot sizes, just-in-time production, and inventory control systems.

Example 5.4 A low-volume machine shop conducted an employee survey, asking for suggestions about where the major opportunities for improvement were. The suggestion that was mentioned most often was "boring operations plantwide." The scrap and rework report also supported the employees' opinion. As the discussion continued, these employee concerns were mentioned:

"There are just so many part numbers that go through the boring department, it will be like trying to eat an elephant." (The problem is just too big and complicated.)

"It is not fair to record bore sizes from part numbers with greatly different tolerances." (Tolerances ranged from ±0.0005 up to ±0.010 in.) "Some parts have a bore size as small as 0.500 in, while others have bores up to over 60 in."

A decision was made to group part numbers into families depending on tolerance and bore size and have one control chart for each family.

Chart no.	Bore size (in)	Tolerance (in)
1	0.500 – 2.000	±0.0005 – ±0.0015
2	0.500 – 2.000	±0.002 – ±0.005
3	2.001 – 5.000	±0.0005 – ±0.0015
4	2.001 – 5.000	±0.002 – ±0.005
5	5.001 – 12.000	±0.0005 – ±0.0015
6	5.001 – 12.000	±0.002 – ±0.005
7	12.001 – 30.000	±0.0005 – ±0.0015
8	12.001 – 30.000	±0.002 – ±0.005
9	12.001 – 30.000	±0.0051 – ±0.010
10	30.000 – 60.000	±0.0005 – ±0.0015
11	30.000 – 60.000	±0.002 – ±0.005
12	30.000 – 60.000	±0.0051 – ±0.010

Further investigation revealed that charts 5 through 9 contained the part numbers with the highest volume of scrap, rework, and deviation requests. All the workers, supervisors, maintenance people, and engineers were informed of the few vital families of part numbers to concentrate their efforts upon. One difficult agreement was reached when the production machine operators and the supervisors all agreed to *aim the process* (tooling and machine) at the middle of the specification (specification nominal). In the past there was a tendency to run the bore on the low side of the specification. An undersize bore would be reworked, but an oversized bore would likely be scrapped or at best require deviation approval. Each of the boring machines had a notebook with 12 sets of control charts.

Machine 708, Chart 8

Part no.	Specification and tolerance (in)	Zero reference point (in)
407-J	18.500 ± 0.003	18.500
1295-J	24.875 ± 0.003	24.875
3097-J	20.000 ± 0.003	20.000

Production rates are rather low (2 to 10 parts per shift). It was decided to use an individual and moving range control chart. It is very important to mark on the chart what part number is being machined when machine changeovers take place and when process changes are observed.

The measuring instrument (dial indicator) is accurate to within 0.0001 in. The target dimension will always be the midpoint of the specification (nominal). The measurements will be coded on the individual chart from a reference point of zero (specification nominal).

In this example the nominal (midpoint) is the target, and we simply plot the individual readings from the reference point of nominal in increments of 0.0001 in.

The combination of part 407-J and machine 708 contributed in excess of $15,000 in scrap, rework, and processing of deviations. There was the impression that the machines were simply not capable. Past scrap reports and statisti-

Machine 708

Part no.	Nominal	Actual reading	Plotted reading	Moving range
407-J	18.500	18.5012	+12	
407-J	18.500	18.4995	−5	17
407-J	18.500	18.4997	−3	2
407-J	18.500	18.5012	+12	15
407-J	18.500	18.4994	−6	18

cal summary sheets indicated that almost 30 percent of the product was produced out of specification. Much of it was undersized and could be reworked. There was a noticeable improvement with just the agreement from the machining department to target the process at nominal.

Figure 5.19 is the individual and moving range chart for part 407-J and part 1295-J. The calculations for the control chart are shown below.

summary In this example, the following tools were used: employees' survey, Pareto analysis, brainstorming, matrices, and finally an individual and moving range control chart. The control chart shows that the process is not in a state of statistical control—the process is not stable. There is out-of-specification prod-

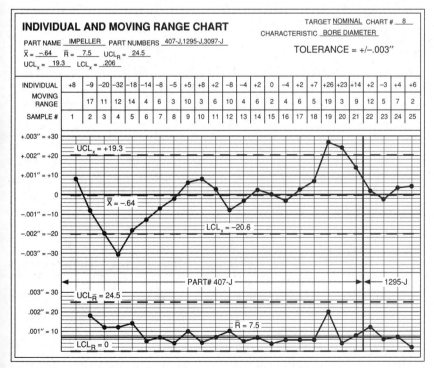

Figure 5.19 Control chart for low-volume manufacturing applications.

uct being produced. The fourth reading (-32) is out of control, outside the lower specification, and outside the control limit. The alarm has gone off, but unfortunately out-of-specification product has been produced.

$$\overline{X} = \frac{\Sigma X}{k} = -\frac{16}{25} = -0.64 = -0.000064 \text{ in below nominal}$$

$$\overline{R} = \frac{\Sigma R}{k-1} = \frac{180}{24} = 7.5 = 0.00075 \text{ in}$$

$$\text{UCL}_X = \overline{X} + E_2\overline{R} = -0.64 + 2.660(7.5) = 19.3 = 0.0019 \text{ in}$$

$$\text{LCL}_X = \overline{X} - E_2\overline{R} = -0.64 - 2.660(7.5) = -20.6 \text{ in}$$

$$\text{UCL}_R = D_4\overline{R} = 3.267(7.5) = 24.5 = 0.0025$$

$$\text{LCL}_R = D_3\overline{R} = 0(7.5) = 0$$

The 19th and 20th measurements are out of control on the high side of the individual chart; both of the individual measurements are within the specification. The control limits have drawn attention to an unstable situation before out-of-specification product is produced. This is the ideal situation. The difficult task ahead is to find the cause of these out-of-control conditions and prevent their recurrence.

5.15 Summary

The purpose of Shewhart control charts for variable quality characteristics is to determine if the output of a process is stable, consistent, and predictable. Control charts do not determine if engineering specifications are being consistently met. The variation observed in a process can be divided into two categories:

1. *Common-cause variation.* This is inherent in the process as it currently exists. It will take action from the management of the company if this type of variation needs to be reduced. The local workforce very seldom has the responsibility to do what is necessary to reduce common-cause variation.

2. *Special-cause variation.* This type of variation is not part of the intended process. The control chart will almost always highlight this type of variation. This type of variation indicates that the process is unstable, inconsistent, and not predictable. The local company workers should investigate the cause of special-cause variation. If the local workforce can take action to prevent this special-cause variation from happening in the future then they should do so. Usually the management team will need to take action to prevent special-cause variation from happening in the future.

- Identify processes that have low levels of customer satisfaction, poor capability, and out-of-control conditions. The long-term strategy should be to bring the processes into a state of statistical control. If the process is close to the customer's target, the level of customer satisfaction will improve when the process is brought into statistical control. There is the chance that the level of customer satisfaction still may not be acceptable. Then we must strive to improve the process even more, which will improve the capability of the process and the levels of customer satisfaction.

- Improve processes so they are functioning in a state of statistical control.
- Have good capability (C_p and C_{pk} of 1.67 minimum).
- Verify that customer satisfaction levels are high.
- Recalculate control limits once a process has been improved. These are signs that a process has been improved:
  1. The process has been adjusted closer to the target.
  2. There is reduced variation in the process (ranges get smaller).
  3. A higher number of the sample statistics fall inside the calculated control limits.
- The control limits are $\pm 3\sigma$ away from the centerline of the chart. If a process is in a perfect state of statistical control then 99.7 percent of the points would be inside the calculated control limits.
- Try to use the instant time sampling scheme if possible. Use overtime sampling if there is very little (if any) variation observed using the instant time sampling method.
- Consider using the individual and moving range chart or the moving average and moving range chart for processes that have homogeneous output.
- If there is a desire to monitor a process that runs very well (high customer satisfaction, good capability and operating in a state of statistical control), consider using a histogram, a run chart, a control chart with modified control limits, a precontrol chart, or some type of checksheet. These techniques will not identify special-cause variation or drive a process toward continual improvement, but these techniques can be used to monitor a process. They are simple to understand and are not very restrictive.
- If the production rate is very low or the lot size is quite small, a charting of the process is the proper scheme.

TABLE 5.1 Shewhart Control Chart Factors

Subgroup size n	Average and Range Chart			
	A_2	d_2	D_3	D_4
2	1.880	1.128	0.0	3.267
3	1.023	1.693	0.0	2.574
4	0.729	2.059	0.0	2.282
5	0.577	2.326	0.0	2.114
6	0.483	2.534	0.0	2.004
7	0.419	2.704	0.076	1.924

Subgroup size n	Medians and Ranges			
	$\widetilde{A}_2$	d_2	D_3	D_4
2	1.880	1.128	0.0	3.267
3	1.187	1.693	0.0	2.574
4	0.796	2.059	0.0	2.282
5	0.691	2.326	0.0	2.114
6	0.548	2.534	0.0	2.004
7	0.508	2.704	0.076	1.924

Subgroup size n	Individuals and Moving Ranges			
	E_2	d_2	D_3	D_4
2	2.660	1.128	0.0	3.267
3	1.772	1.693	0.0	2.574
4	1.457	2.059	0.0	2.282
5	1.290	2.326	0.0	2.114
6	1.184	2.534	0.0	2.004
7	1.109	2.704	0.076	1.924

Chapter 6

Capability

Process capability is the second phase of evaluating a process. Once the process shows a good state of statistical control, the next step is to determine the capability of the process to meet the specification requirements.

6.1 Principle of Capability

Most manufacturing operations calculate the capability values for their products and processes to answer the following questions:

1. Will the distribution of individual measurements easily fit between the specification limits? The C_p value answers this question.
2. Is the process centered adequately, so that neither tail of the distribution of individuals is close to being outside the specifications? The C_{pk} value answers this question.

The two most common indicators of process capability are the C_p index and the C_{pk} index. The C stands for capability; the p stands for process, and the k stands for the index.

6.2 The C_p Index

The C_p value is occasionally referred to as the *process potential*. The formula is

$$C_p =$$

where $\hat{\sigma}$ = estimate of standard deviation of individuals made by using $\hat{\sigma} = -R/d_2$. Remember that the great majority of the normal distribution (99.7 percent) is within $\pm 3\sigma$ from the center of the

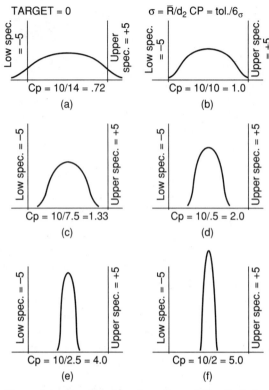

Figure 6.1 Graphical explanation of the C_p index. Process must be in control before capability can be calculated accurately.

process. The C_p index is nothing more than a ratio of the specification width to the width of the normal distribution.

The formula used to calculate a standard deviation of individual measurements (x's) is

$$\hat{\sigma} = \frac{\bar{R}}{d_2}$$

Figure 6.1 graphically shows the concept of C_p statistics. Notice that in all six of the situations (a) through (f) in Fig. 6.1, the process is perfectly centered at the midpoint of the specifications. In all six situations shown, the tolerance is 10 units wide. The distribution of individuals is varying in dispersion or spread. In Fig. 6.1(a), the distribution is 14 units wide. Unfortunately, the specification is only 10 units wide. This process is not capable of easily meeting the specification of ±5 units. In Fig. 6.1(a), there is a small portion of the tails of the distribution out of specification on both the high and low ends.

In Fig. 6.1(b), the tolerance width and the $6\hat{\sigma}$ distribution of individuals are the same width. The specifications in Fig. 6.1(b) are $\pm 3\hat{\sigma}$ from the center of the process. Theory tells us that 99.7 percent of the individual measurements are between the specifications. In Fig. 6.1(b) the process is just barely meeting specifications.

Figure 6.1(c) shows that the specifications are $\pm 4\hat{\sigma}$ from the center of the process. Theory tells us that 99.994 percent of the individual measurements are within the specification limits.

The process graphically described in Fig. 6.1(f) has a C_p value of 5.0. Five normal distribution curves could fit inside the specification limits. The tolerances in Fig. 6.1(f) are $\pm 15\hat{\sigma}$ from each other.

QUESTION: Are there not some situations where all the output from the process in Fig. 6.1(a) will function well and have high levels of customer satisfaction even though some of the product is out of specification?

ANSWER: I am confident that there are some situations where this is true. If that is the situation, then what does the management of the company tell the worker on the factory floor? The direction from management could be something like this. "Even though the specification is ± 5 units, as long as the product is between ± 6 units, keep the machine running and have the products sent to the next operation." Or the directions to the worker could be the following. "The engineering department has approved a waiver which allows a wider specification than the original ± 5, so we can keep the machine running."

If the entire output in Fig. 6.1(a) functions well in assembly, test, and field operations, then obviously the functional limits of the product being produced are not the specification limits. Another option is for the department responsible for setting specifications to revise the specification of ± 5 units to what the process actually needs. The workers from the factory floor should not be subjected on a daily basis to operating equipment that produces out-of-specification product. Worker morale will deteriorate. There will be confusion as to the quality policy of the company. Engineering requirements should be reviewed at the early stages of product development. Manufacturing processes should have proven capability before the processes are released to the production departments.

QUESTION: Is the process in Fig. 6.1(f) not overkill, spending a dollar to save a dime?

ANSWER: I doubt if the process in Fig. 6.1(f) is one of the major opportunities for improvement in quality, productivity, or customer satisfaction. Many companies have processes similar to the process in Fig. 6.1(f). Processes similar to that shown in Fig. 6.1(f) with excellent capability just do not get much attention. The big opportunities for improvement are probably processes with capability similar to Fig. 6.1(a), (b), and possibly (c); these are the processes that we should concentrate our efforts on. If a company had few processes similar to Fig. 6.1(a), (b), or (c), but many processes like Fig. 6.1(e), or possibly Fig. 6.1(f), then it might make sense to find a way to reduce manufacturing processing time and costs or material costs. Consider the following:

Process in:	Capability	Total cost* ($)
Fig. 6.1(d)	2.0	10.00
Fig. 6.1(e)	4.0	10.50
Fig. 6.1(f)	5.0	10.75

*Total cost includes all manufacturers' costs as well as direct, indirect, and material costs.

If quality is defined as making product with little variation around the target, then the process in Fig. 6.1(f) has the highest quality. Today, competitive customers want high quality at low cost along with timely product delivery.

Most companies that use problem-solving techniques have a written policy as to their goal or minimum capability requirement for critical processes. Figure 6.2 shows how we should think of capability. If $C_p < 1.0$, then the process is classified as not capable and requires management attention to improve the situation.

The goal that most companies are striving for is a C_p number of 1.67 minimum. It is not correct to think of a process that has a C_p value of 1.68 as acceptable and a process with a C_p value of 1.65 as not acceptable. There is really not much difference in the quality level from two processes, one with a C_p value of 1.65 and another with a C_p value of 1.68. As with many things in life, there are "many shades of gray."

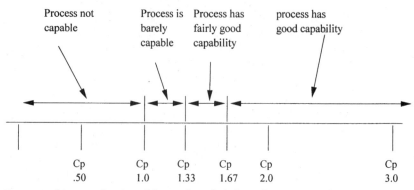

Figure 6.2 Line graph categorizing various C_p index values.

6.3 The C_{pk} Index

The second statistic that is used to evaluate capability is known as the C_{pk} index number. The C_{pk} index number answers the following question: From the center of the true process ($\overline{\overline{X}}$), how many standard deviations both forward and backward are the nearest of the two specifications?

In short summary, C_{pk} tells us whether the process is targeted adequately vis-à-vis the specifications. This applies to finished characteristics and dimensions as well as manufacturing process parameters.

If $C_{pk}<1.0$, this tells us that some portion of one or both tails of the distribution of individual measurement are beyond one or both of the specification limits.

There are a number of ways of calculating the C_{pk} value. The first set of formulas is as follows:

$$Z_{\text{LSL}} = \frac{\overline{\overline{X}} - \text{LSL}}{\hat{\sigma}} \tag{6.1}$$

$$Z_{\text{USL}} = \frac{\text{USL} - \overline{\overline{X}}}{\hat{\sigma}} \tag{6.2}$$

The next step is to determine which of the two calculations represents the worse condition, which of the two specifications is closer to the middle of the distribution. The specification that is closer to the center of the process will be the smaller of the two Z values calculated. We now refer to that number as *Z minimum*. The last step in the calculation for C_{pk} is as follows:

$$C_{pk} = \frac{Z \text{ minimum}}{3}$$

Note: The 3 in the denominator of the above formula represents half of the number of standard deviations σ that the normal distribution is wide.

When the process is perfectly centered on the middle of the specification, the C_p value and the C_{pk} value will be exactly the same. Most companies' goal is for C_{pk}, as with C_p, to have a C_{pk} of 1.67 minimum. Some companies have a capability goal of 1.33 minimum.

When control charts are used to improve a process, three goals must be attained before the process can be considered acceptable:

1. In control
2. Minimum C_p value of 1.67
3. Minimum C_{pk} value of 1.67

When these three goals have been attained, the problems associated with this process are usually trivial in magnitude. The most important and the most difficult goal to attain is the goal of *statistical control*.

Example 6.1 An average and range chart is being used to monitor a machining process for a nominal-is-best (NIB) quality characteristic. The process shows a good state of statistical control. The capability will be calculated based upon the following information from the control chart:

$$\overline{\overline{X}} = 0.5071 \text{ in} \qquad \overline{R} = 0.0052 \text{ in}$$
$$\text{USL} = 0.520 \text{ in} \qquad \text{LSL} = 0.500 \text{ in}$$
$$\text{Target} = 0.510 \text{ in}$$
$$n = 4 \qquad d_2 = 2.059$$

$$\hat{\sigma} = \frac{\overline{R}}{d_2} = \frac{0.0052}{2.059} = 0.00252 \text{ in} \tag{6.3}$$

$$C_p = \frac{\text{tolerance}}{6\hat{\sigma}} = \frac{0.020}{0.01512} = 1.32 \tag{6.4}$$

The C_p value of 1.32 means that almost 1⅓ distributions will fit between the specifications. The next step is to calculate the C_{pk} value.

$$Z_{\text{LSL}} = \frac{\overline{\overline{X}} - \text{LSL}}{\hat{\sigma}} = \frac{0.5071 - 0.500}{0.00252} = 2.82 \tag{6.5}$$

$$Z_{\text{USL}} = \frac{\text{USL} - \overline{\overline{X}}}{\hat{\sigma}} = \frac{0.520 - 0.5071}{0.00252} = 5.12 \tag{6.6}$$

$$C_{pk} = \frac{Z_{\min}}{3} = \frac{2.82}{3} = 0.94 \tag{6.7}$$

Figure 6.3 graphically shows the capability of the process.

The low specification of 0.500 in is 2.82 standard deviations from the center of the process. The high specification is 5.12 standard deviations from the cen-

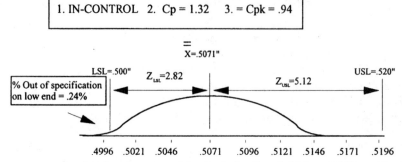

Figure 6.3 Graphical explanation of the C_{pk} index (board thickness) NIB quality characteristic.

ter of the process. The worst condition is with the low specification. The C_{pk} of 0.94 tells us the following: The nearest specification is 94 percent of 3 standard deviations from the center of the process.

The action to take now is to center the process closer to the target of 0.510 in. The big gains are not in reducing the percentage of out-of-specification product. The big gain is achieved by shifting the entire process closer to the target of 0.510 in. More will be gained both financially and in productivity by striving to improve the processes to the extent that they will operate in a state of statistical control (in addition to statistical control it is very important that the process and product characteristics be centered close to the target).

The calculations of C_p and C_{pk} both are based upon the assumption that the distribution of individuals is the shape of a normal curve. Both formulas use standard deviation in the calculations, and standard deviation is based upon the distribution being normally distributed.

> QUESTION: Since the output of some processes does not take the shape of the normal distribution, does this mean that you cannot calculate the capability or that if you do calculate the capability, it is of no value?
>
> ANSWER: Not necessarily. This does not mean that the calculations for capability are all for naught. Recall the discussion earlier explaining the normal distribution. Remember that 68 percent of the measurements are between $+1\sigma$ and -1σ and that 95 percent of the measurements are between $+2\sigma$ and -2σ. If the shape of the distribution of individuals differs greatly from the normal distribution, then we should not expect 68 percent of the individual readings to be within $\pm 1\sigma$; also do not expect that 95 percent of the individuals will actually be within $\pm 2\sigma$. The normal probability paper method can be used to determine the capability of processes that are not the shape of the normal curve. This method will be shown later in this chapter.

Still, you should work to attain the three goals when you use control charts:

1. A state of statistical control
2. Minimum C_p value of 1.67
3. Minimum C_{pk} value of 1.67

Example 6.2: Oven temperature (process control)

Target = 340°F $\bar{\bar{X}}$ = 346°F $\bar{R}$ = 7.2°F d_2 = 2.326

In this example, there are no actual process parameter limits (specifications), but the technical people at this company agree that oven temperature is a critical process parameter, and they are confident that an oven temperature of 340°F is best.

The first thing to do is to compare the target temperature (340°F) with the average (346°F). It is obvious that the oven temperature is running off target

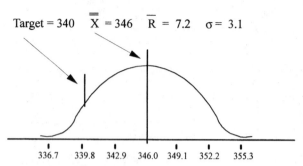

Figure 6.4 Oven temperature capability (process parameter, no actual specification exists).

on the high side of 6°F. The second step is to determine how widely the distribution of temperature is varying. (We want to find out how wide the normal distribution of oven temperature is.)

The formula to use is

$$\hat{\sigma} = \frac{\overline{R}}{d_2} = \frac{7.2}{2.326} = 3.1°F$$

In past discussions, we have talked as though the normal distribution were 6 standard deviations wide, so in this case $6\sigma = 18.6°F$. Another way of saying this is that the oven temperature can be held to $\pm 9°F$. Visually this is shown in Fig. 6.4.

The discussion of this situation would probably be something like this:

> We should adjust the oven temperature closer to the target temperature of 340°F. The other item for discussion is whether or not the company can tolerate a temperature variation of 18°F. If the answer is yes, we can tolerate that amount of variation, then we should continue to monitor the process, keeping the oven temperature on target and in statistical control. If we decide that we cannot tolerate that much variation, then we must go through some brainstorming or problem-solving activities to help us determine what to change in the process that will reduce this variation.

Every now and then there will be someone who will persist in requiring that you provide your customer with the C_p and the C_{pk} numbers for the process. We must determine what specification limits will give us the desired C_p and C_{pk} values. From earlier discussion we recall that a value of 1.67 for C_p and C_{pk} means that the tolerances are 10 standard deviations from each other. Recall that earlier in this example the standard deviation was 3.1°F, so 5 standard deviations would be 15.5°F. So in this case we would probably request a specification limit of $340 \pm 15°F$.

This does not mean that the specification department will agree to the tolerance of $340 \pm 15°F$, but this is the specification that the current process needs so that the capability of the process has C_p and C_{pk} values of 1.67.

Capability ratio (C_r)

A number of years ago some companies used the *capability ratio* C_r instead of the C_p index. The formula for the capability ratio is

$$C_r = \frac{6\hat{\sigma}}{\text{tolerance}}$$

This formula is the reciprocal of the formula for C_p'. For example,

$$C_r = \frac{1}{C_p}$$

$$C_p = 1.67 \quad C_r = \frac{1}{1.67} = 0.60$$

Since the capability ratio C_r is the reciprocal of C_p, the equivalent goals are

1. Process in control
2. A maximum C_r of 0.60 (60 percent)
3. C_{pk} value of 1.67

Not many companies use the capability ratio in place of the C_p index anymore.

6.4 Larger-Is-Better Characteristics

For either larger-is-better (LIB) or smaller-is-better (SIB) characteristics, the C_p value is not normally calculated. Usually only C_{pk} is calculated. Example 6.3 shows the reasoning for this. Up until this point, first C_p was determined. The C_p index answered the question, What is the relationship of the specification width to the width of the normal distribution of individual measurements?

Example 6.3

Specification = 42,000 lb/in² minimum tensile strength

Average = 46,350 lb/in² $\hat{\sigma} = 730$

How wide is the tolerance? Is it from 42,000 lb/in² to an infinitely high tensile strength?

The capability index C_{pk} will give us the information we are interested in. There is a modified formula for both LIB and SIB types of quality characteristics. For example, the strength of a material is specified as a minimum requirement 42,000 lb/in². There is only one specification, so therefore only one Z value will be calculated.

$$Z_{SL} = \frac{\overline{\overline{X}} - SL}{\hat{\sigma}}$$

where SL = specification limit. The last calculation is

$$C_{pk} = \frac{Z_{SL}}{3}$$

Target = ____ average = 46,350   sigma = 730   specification = 42,000 psi min

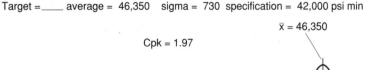
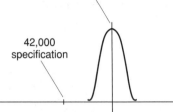

Figure 6.5 Tensile strength capability (minimum strength specified—C_{pk} is primary concern).

The goal for capability still applies for these types of characteristics. A C_{pk} of 1.67 minimum is common. In this example the C_{pk} value is

$$C_{pk} = \frac{\overline{\overline{X}} - \text{SL}}{3\hat{\sigma}} = \frac{46{,}350 - 42{,}000}{2190} = 1.97$$

The specification is 197 percent of 3 standard deviations from the center of the process. The capability would be considered quite good. Figure 6.5 shows the capability of the tensile strength.

QUESTION: In the past, C_{pk} was calculated differently. The Z_{LSL} and Z_{USL} were calculated first; then the smaller of those two Z values was divided by 3, and that answer was C_{pk}. Using this method, you cannot determine the proportion out of specification. Should not the Z for this situation be calculated first, so the proportion out of specification can be determined from the Z table?

ANSWER: You have raised a good point. Using the method just explained, you cannot determine the proportion out of specification. To get the information you were concerned about, the problem could be worked in this way:

$$Z_{\text{SL}} = \frac{\overline{X} - \text{SL}}{\hat{\sigma}} = \frac{46{,}350 - 42{,}000}{730} = 5.95$$

$$C_{pk} = \frac{Z_{\text{SL}}}{3} = \frac{5.95}{3} = 1.98$$

In this example, there is a negligible proportion out of specification. The Z table in this book only goes up to 4.00.

6.5 Smaller-Is-Better Characteristics

For SIB type of characteristics, the most common formulas are

$$Z_{\text{SL}} = \frac{\text{SL} - \overline{X}}{\hat{\sigma}}$$

Average = 23.7 cc   Standard deviation = 5.38 cc   Specification = 30 cc
maximum

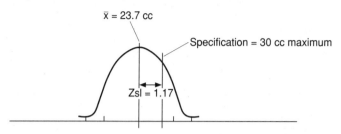

Figure 6.6 Process capability of leak rate (maximum specification specified).

$$C_{pk} = \frac{Z_{SL}}{3}$$

Example 6.4: Leak rate (maximum specified) The specification for a sealed assembly is 30 cm³/min at 20 lb/in². From a moving average and moving range chart with a sample size of 3, we find

Average = 23.7 cm³   Average range = 9.1 cm³ $d_2 = 1.693$

The estimate for the standard deviation must be calculated first.

$$\hat{\sigma} = \frac{\bar{R}}{d_2} = \frac{9.1}{1.693} = 5.38 \text{ cm}^3$$

Next we must determine the Z value for the specification limit, using the formula shown here:

$$Z_{SL} = \frac{SL - \bar{X}}{\hat{\sigma}} = \frac{30 - 23.7}{5.38} = 1.17$$

Figure 6.6 shows that the specification limit of 30 cm³ is up 1.17 standard deviations from the center of the process. In this case there is a sizable proportion of assemblies (12.1 percent) above the specification limit of 30 cm³. The C_{pk} value is going to be bad.

$$C_{pk} = \frac{Z_{SL}}{3} = \frac{1.17}{3} = 0.39$$

QUESTION: These calculations tell us that we are producing 12.1 percent of product beyond the specification of 30 cm³. What is the benefit of these capability calculations?

ANSWER: These calculations tell us quite a bit about the capability of the process, but they do *not* improve a process, reduce the rejection rate, or improve customer satisfaction. The idea we are explaining is the capability. The cornerstone of determining capability is that the process shows a good state of statistical control. In the past, many times I heard comments from people stating that many manufacturing processes are not capable of meeting specification requirements. What these people are really saying is that they are having trouble with many of their processes meeting the specifications. Usually people making these types of comments do not have knowledge about the state of statistical control of their processes.

In Example 6.4, what has been shown is that the current process cannot meet the specification of 30 cm^3/min. The next steps in that example would be as follows:

1. Identify where in the assembly the leak is occurring.
2. Brainstorm a list of factors that would cause the assembly to leak.
3. Gather data relating to the factors causing the leak (piece part dimensions, flatness, smoothness, bolt torque, type of assembly method, etc.).
4. Work on improving the elements in the process that show a need for improvement based upon the data gathered.
5. Consider conducting some type of multifactor designed experiment to find the optimum combinations of process parameters and product design parameters.

QUESTION: Is it acceptable to calculate the capability if there is no control chart that shows the process in a state of control? Suppose that I have 100 measurements. The average and standard deviation could be calculated by using the method in Chap. 3. To determine capability, all that is needed is the average, standard deviations, and the specifications. I think many people will not have the patience to wait for the process to be brought into a state of control. The topic of capability holds great interest for many of the people that I have talked to.

ANSWER: This would be a very bad practice to follow. The concept of process capability assumes that the process shows a good state of statistical control. Recall that the first goal is statistical control. If a company is interested in improving the process so that the process is running in prevention mode, then the main focus should be to bring the process into a state of control. I know that it takes much patience to "stay the course" of process improvement. If there is great interest in the extent to which the process is meeting the specification, then go ahead and calculate the average and standard deviation and conduct the calculations for capability. Do not refer to the values calculated as the C_p or the C_{pk} value of the process. On the data sheet or report, inform all the parties that the capability of the process is not known yet, since the process is not in a state of control and does not have good stability.

6.6 Using Normal Probability Paper to Calculate Capability

The standard methods for determining the capability of a process should not be used if the process differs greatly from the shape of the normal distribution. Normal probability paper should be used.

Example 6.5 One of the examples in Chap. 5 involved the evaluation of the impurity level for a cosmetic product (see Fig. 5.16). The 25 individual measurements from Fig. 5.16 have been grouped into cells to aid in filling out the normal probability paper (Fig. 6.7).

Lower boundary	Upper boundary	Cell midpoint	Frequency
0	9.9	5.0	0
10.0	19.9	15.0	0
20.0	29.9	25.0	4
30.0	39.9	35.0	6
40.0	49.9	45.0	7
50.0	59.9	55.0	1
60.0	69.9	65.0	4
70.0	79.9	75.0	1
80.0	89.9	85.0	2

The method for completing the form was explained in Chap. 3.

Transfer the cell midpoint values and the frequency to the graph paper.
Calculate the *estimated accumulated frequency* (EAF).
Calculate the plot points.
Transfer all the plot point values onto the graph portion of the paper.
Connect all the plot points.
Recall that the specification is 60 parts per million (ppm) maximum.
Line up where 60 ppm is located on the bottom scale; move up until the line is reached.
Move to the right until you reach the scale.

Visual analysis tells us that approximately 26 percent of the product is over the specification.

Finally we should determine the C_{pk} value. In this situation the formula shown here should be used.

$$\hat{C}_{pk} = \frac{\text{SL} - X_{50}}{X_{0.135} - X_{50}}$$

$$= \frac{60 - 44}{108 - 44} = \frac{16}{64} = 0.25$$

where SL = specification limit
X_{50} = 50th percentile value
$X_{0.135}$ = 0.135 percentile value (equivalent to $+3\sigma$)

156   Chapter Six

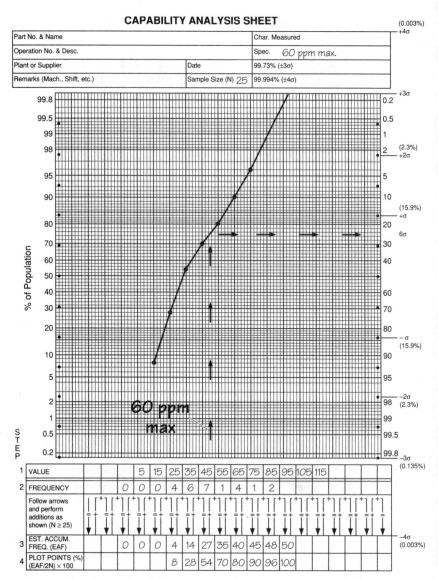

Figure 6.7 Use of normal probability paper to determine process capability.

In this case C_{pk} should be reported as 0.25, which is very bad.

This method for determining capability is much more cumbersome, but it is more mathematically correct.

6.7 Summary

1. Capability for processes should be calculated only after a process shows a good state of statistical control.

2. The C_p index evaluates the process to see if the process has the potential to meet the specification. The C_p index does not evaluate the targeting of the process.

3. The C_{pk} index evaluates the actual performance of the process. This index does address the targeting of the process.

4. If the process is perfectly centered in the middle of the specification, C_p and C_{pk} will be the same.

5. Some companies are concerned about only the C_{pk} index, because if the C_{pk} index is acceptable, then the C_p index will also be acceptable.

6. Most companies have capability goals of 1.67 minimum for both C_p and C_{pk}.

7. None of the calculations in this book by themselves will improve quality, save the company money, increase productivity, or improve customer satisfaction. It is just not that simple. Process improvement usually is accomplished little bit by little bit. The upper management of the company must lead the workforce by giving guidance, support, and time for the results of process improvement to materialize for this to be a beneficial activity.

6.8 Problems

Calculate the C_p and C_{pk} capability indices for the following examples. Write a short two or three sentence summary of your analysis of the capability of each of the processes.

1. $\bar{\bar{X}} = 30.71 R_C$; $\bar{R} = 2.1 R_C$; USL = $34 R_C$; LSL = $26 R_C$;
   $n = 4$; $d_2 = 2.059$

2. $\bar{\bar{X}} = 4.3766$ in; $R = 0.00254$ in; USL = 4.365 in; LSL = 4.385 in;
   $n = 3$; $d_2 = 1.693$

3. $\bar{\bar{X}} = 46.7$ ft·lb; $\bar{R} = 4.6$ ft·lb; USL = 50 ft·lb; LSL = 40 ft·lb;
   $n = 4$; $d_2 = 2.059$

4. $\bar{\bar{X}} = 8.1439$ in; $\bar{R} = 0.0021$ in; USL = 8.154 in; LSL = 8.134 in;
   $n = 3$; $d_2 = 1.693$

Chapter 7

Variable Control Charts, Multivary Charting, and Precontrol Charts

7.1 Recording Process Changes on Control Charts

If you ask representatives from 100 companies why they are using control charts or other tools, most likely the response will be one of the three following reasons:

1. "The only reason I am using these charts is because the customer is requiring me to use them."
2. "The reason I am using these charts is to keep an eye on the process."
3. "We use these control charts as a tool to help us improve our processes."

Of course, the reason that Dr. Deming encourages companies to use the tools of statistical problem solving is to improve processes (response 3). There are numerous nonstatistical techniques that will help improve a process. One of the most overlooked points that should be stressed more during employee training is the importance of recording notes or comments on the control chart or data collection sheet. These notes explain process changes made or factors investigated that are causing the out-of-control conditions.

Many people improperly think that the main objective is for the operator to "keep" a process in statistical control. The real objective should be for the *management* of the company to improve a process so that the process will stay in a state of statistical control. Improving a process to the extent that the process stays in perfect statistical control is a never-ending challenge. There will be changes in the elements of the process that cause out-of-specification conditions which are part of normal operating conditions. The challenge is to minimize the frequency and severity of these abnormal fluctuations.

You may ask, How do I learn to improve the process so that it will stay in control? The answer is to gain knowledge from notes or comments on past control charts, then analyze and prioritize that information. Finally, take action to improve the process by preventing the past causes of out-of-control conditions from happening in the future.

When operating personnel are being taught to fill out control charts, often the importance of recording notes and comments on the control chart is not stressed. Some people are hesitant for numerous reasons to record information on the chart. One way to minimize this hesitancy is to develop a *menu* of predetermined causes so that the operator will record only codes that mean specific process changes or adjustment. For example, in a machining operation some process changes or adjustments could be

1. Tool change
2. Tool adjustment
3. New lot of material
4. Feed rate adjustment
5. Tool speed adjustment
6. Spindle realignment

In an injection molding operation, these are some possible process changes or adjustments:

1. New lot of material
2. Increase in injection pressure
3. Decrease in injection pressure
4. Increase in mold temperature
5. Decrease in mold temperature
6. Increase in injection time
7. Decrease in injection time
8. Increase in clamp pressure
9. Decrease in clamp pressure

The person in charge of the process just writes in the code on the chart. Sometimes further explanation will be needed beyond just the code. The objective is to minimize the reluctance of the operating personnel to document process changes or causes of out-of-control conditions so that the changes can be analyzed. Cause-and-effect relationships can be investigated so that the process will come closer to the goal running in a state of statistical control. This list of menu items can be developed prior to releasing a control chart or another type of statistical tool to the production floor. The list of menu items should be generated by a cross-functional team with representatives from the process engineer, quality engineer, machine operators, and supervisors.

7.2 Visual Pattern Analysis of Control Charts

For a process to be in a state of perfect statistical control, the great majority (99.7 percent) of the sample statistics must be inside the control limits and there will be no unnatural pattern of variations. This section of the book shows the more common and important unnatural patterns of variations that should be familiar to all the parties using these control charts. Sometimes people are under the impression that their processes are in statistical control when they are actually out of control. This can cause much confusion and misdirected effort.

The following pages show some common mistakes made in visual pattern analysis. Figure 7.1 lists some of the requirements for a

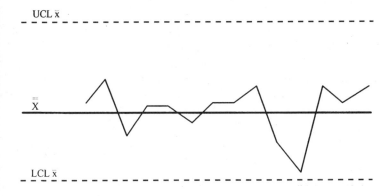

For a process to be in a perfect state of statistical control all of the following must be observed:
1. Most of the points (about 68%) must be near the centerline of the control chart
2. A few (about 5%) of the points must be near the control limits
3. The great majority of the points must be inside the control limits

Figure 7.1 Pattern requirements for control charts.

process to be considered in a state of statistical control. For a process to be in a perfect state of statistical control, these three requirements must be observed:

1. Most of the points (about 68 percent) must be near the centerline of the control chart. This requirement should be scrutinized strongly for an average or a median chart. The enforcement of this requirement should be relaxed somewhat for an individual chart or the range chart. The distribution of individuals and/or ranges may not be the shape of the normal distribution, so these rules should not be enforced as strongly.

2. Approximately 5 percent of the points should be rather close to the control limits.

3. The great majority of the points on the control chart should be inside the control limits. The control limits are 3 standard deviations away from the centerline, so in theory 99.7 percent of the points on the control chart should be inside the limits when a process is in a perfect state of control.

The rules for zone analysis (Fig. 7.2) apply most for average charts and median charts. The zone analysis for individual charts and range charts should be enforced less strictly. The reason is that averages and medians will conform very closely to the shape of the normal distribution when the process is in statistical control. The distribution of ranges will be the shape of a skewed distribution as long as the sample size is 6 or less.

See Fig. 7.3. This unnatural pattern could be caused by tool wear or the viscosity of a material changing with time. A buildup of material on locating surfaces of a machining jig or fixture could also cause this pattern of trends. This pattern is more commonly seen on the chart that measures central tendency (average, median, or individuals

UCL $\bar{x}$		$+3\sigma \bar{x}$
	2%	$+2\sigma \bar{x}$
	14%	$+1\sigma \bar{x}$
$\bar{\bar{X}}$	34%	
	34%	$-1\sigma \bar{x}$
	14%	$-2\sigma \bar{x}$
LCL $\bar{x}$	2%	$-3\sigma \bar{x}$

Figure 7.2 Zone analysis requirements for control charts.

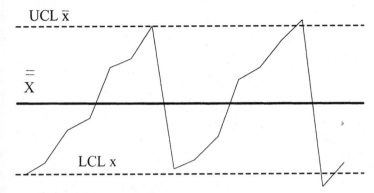

Figure 7.3 Pattern analysis for trends.

chart). Some processes naturally show this type of variation, but still it is an unwanted pattern. The challenge is to find a way to modify the process so that there is less of a slope. If the slope of the process can be reduced, some advantages are (1) that the production runs before the process must be adjusted are longer and (2) the process can be set initially closer to the target and not allowed to trend up as long as was done previously.

An upward trend on the range chart is an indication that the variation in the process is increasing. Some possible causes could be the increased wear on one part of a machine or that some element in the process has changed significantly and the current level of that process factor is causing the variation in the process to increase. If a downward trend is observed, the cause of reduced variation should be discovered and made a part of the process.

In Fig. 7.4, there are too many points out by the control limits and not enough points in toward the centerline of the chart. This pattern

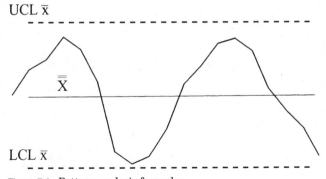

Figure 7.4 Pattern analysis for cycles.

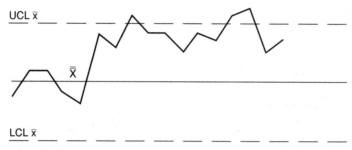

Figure 7.5 Pattern analysis for sudden change in level.

usually means that the shape of the distribution of the individual readings is bimodal. There is not one consistent target. Some causes of cycles are two or more process streams feeding into the process that are targeted differently or two different measuring instruments that are calibrated differently.

There will be great disparity between the estimated capability of the process and its actual output. The level of quality of the output will be much worse than what it is believed to be when one is looking at the calculations for capability.

See Fig. 7.5. This pattern is obvious—the averages are no longer crisscrossing the centerline of the control chart. If this pattern is seen on the average chart, it is an indication that the process has been adjusted (targeted) differently. Sometimes this pattern is referred to as a *run*. This is not always a bad situation. If this change in level is now closer to the target, this is a good situation. It is an unnatural pattern, but it is an unnaturally good situation. The cause of this change in level may be as simple as the factory personnel reacting to the "target" concept. If the process is now centered closer to the target value which is good, all parties involved should be made aware of this change. Put instructions in place to ensure that the process continues to be aimed at the target. The control limits should be recalculated around the process if it is running closer to the target.

A change in the level on the range chart could be either good or bad, depending on whether the change in level is upward or downward. If there is a sudden change upward in the level or ranges, the causes for this bad situation must be determined and corrected.

See Fig. 7.6. There are no points close to the control limits. The most common causes for this unnatural pattern are as follows:

1. The process has improved greatly in the past year or so, but new tighter control limits have never been recalculated.
2. Some type of multistream process is being monitored (multiple fixtures, dies, or cavities). If the streams are aimed differently, the

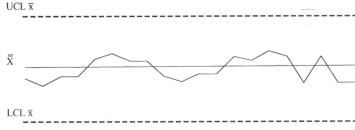

Figure 7.6 Pattern analysis for stratification.

ranges will be exaggerated and will cause the control limits on the average chart to be very far apart.

3. There is a mistake in the calculations for the control limits or in the plotting of the control limits.

7.3 Summarizing Continuous Improvement

> QUESTION: Is there some way to document the gains being made week after week, or month after month, toward the goal of perfect statistical control and acceptable capability? It seems that part of what you have been saying is to just have faith and work at improving the process and things will improve. I think it would be best to have some kind of formal tracking system to determine whether gains are being made.
>
> ANSWER: Yes there are numerous ways to summarize the tracking of process improvement. Some companies want to track the capability information only. The topic of statistical control does not seem to be a concern for some reason. Maybe everyone just assumes that the process is in control and that there is no unnatural variation, so the capability is of primary interest. The form shown in Table 7.1 is used by some companies to track process improvement.

February was the first month of gathering measurements. When there were enough samples, the control limits were calculated to determine whether the process was in control during the past month. The percentages of points inside the control limits fell far short of the 99.7 percent figure for a process in control. Cycles and runs were observed on the average chart. The capability of the process should not be estimated because the process is badly out of control. These limits were drawn in on the control charts to be filled out the following month. At the end of March, the control chart was analyzed. It was found that 84 percent of the averages $\overline{X}$ and 92 percent of the ranges were inside the limits calculated from the first month's measurements. Cycles were observed on the average chart for the month of March. The settings of two process parameters were changed during

TABLE 7.1 Process Improvement Form

Product name_____ Product code_____ Plant_____ page___ of___
　　　　　　　　　　　Specification = 28–36　Target = 32

Time	Percentage of points inside limits		Unnatural patterns	$\bar{\bar{X}}$	$\bar{R}$	C_p	C_{pk}
	Average	Range					
February	76	88	Runs, cycles, $\bar{x}$	34.8	2.3	*	*
March	84	92	Cycles	33.5	2.1	*	*
April	88	88	None	33.2	2.2	*	*
May	92	96	None	32.9	1.9	1.12	0.68†
June	88	92	Cycles	32.3	1.9	*	*
July	96	96	None	32.2	1.7	1.24	1.08

*Process out-of-control capability cannot be accurately determined.
†Process adjusted to target of 32.

March. The grand average $\bar{\bar{X}} = 33.5$ and the average range $\bar{R} = 2.1$ were recalculated with the March measurements. At this time new tighter control limits should be calculated for the control chart to be filled out during April. The combination of the following things is the basis for calculating the new control limits:

1. A higher percentage of points inside the control limits
2. The grand average $\bar{\bar{X}}$ (33.5) getting closer to the target of 32.0
3. The average range $\bar{R}$ decreasing from 2.3 to 2.1
4. Knowledge of which process parameter settings improved the process

The completed chart at the end of April shows about the same percentage of points inside the control limits. Remember, these new control limits are tighter than they were originally. There are no unnatural patterns observed on either chart. The process is closer to the customer's target of 32. The grand average of the readings for April is 33.2. The average range $\bar{R}$ increased from 2.1 to 2.2; this is an undesired event. Larger ranges indicate increased variation in the process. The control limits were not recalculated at the end of April. The control limits for May were kept the same as in April. Corrective action was taken on the process based upon the causes for out-of-control conditions during April.

The review of the control chart at the end of May showed a continued increase in the number of points inside the control limits. The grand average $\bar{\bar{X}}$ for the readings from May (32.9) had shifted closer to the target. The average range $\bar{R}$ has again come down from 2.2 to 1.9. At this time the capability of the process is calculated. The C_p value is 1.12, and C_{pk} is 0.68. Because the process is still not centered as it should be, C_{pk} drops off considerably from the C_p value. The process is barely capable of meeting the specification, even if the process were centered exactly on the target of 32. New tighter control limits were calculated from the grand average and average range of May readings.

During June, the process showed a poorer state of control. Cycles were also observed on the average chart. There was a big improvement in centering the process closer to the target of 32 during June. The grand average $\overline{\overline{X}}$ for June was 32.3. There was no reduction in variation observed during June, and the average range $\overline{R}$ stayed at 1.9.

Tighter control limits were not recalculated, but the control limits for the average chart were lowered 0.6 unit since the process is centered much closer to the customer's target. The process has improved greatly even though there was no reduction in variation.

During July, the company conducted a multifactor designed experiment to determine what factors were variation reduction factors. The grand average $\overline{\overline{X}}$ for July was 32.2. The average range $\overline{R}$ decreased from 1.9 to 1.4. The capability of the process was recalculated. The C_p value was 1.47, and C_{pk} was 1.39. The goal for capability at this company is 1.67 minimum for both C_p and C_{pk}.

The two rightmost columns in the table describe the capability of the process. This seems to be what everyone wants to find out. If the process is not in control, these capability values will be very misleading. If the long-range goal is to have good capability, the logical way to do this is to work for statistical control and center the process close to the customer's target. Even though the capability goals of 1.67 minimum have not been attained yet, there has been great improvement in the process and the quality of the product.

7.4 Median and Range Control Charts

The control chart for medians and ranges ($\widetilde{X}$, R) is very similar to the average and range chart $\overline{X}$, R. The differences are as follows:

1. The sample medians are charted instead of the averages as a measure of central tendency.
2. The central line for the median chart is the average median $\overline{\widetilde{X}}$.
3. A different factor is used in calculating the control limits for the median chart, $\widetilde{A}_2$.
4. The control limits for the median chart are 3 standard deviations of the medians $3\sigma_{\widetilde{X}}$ from the centerline.

Median charts are simpler to use on the factory floor if the educational level of the workforce is low. Figure 7.7 is a median and range

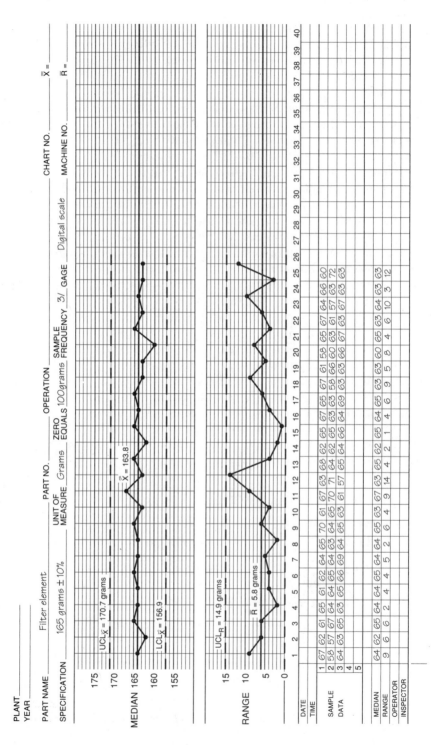

Figure 7.7 Median and range control chart.

chart. The quality characteristic is the weight of the amount of filter element material. The specification is 165 ± 10 percent. The upper specification is 181.5 g, and the lower specification is 148.5 g. Three shots from the filter media dispensing machine are recorded and charted every 30 min.

The calculations for the control limits are shown here. (See Table 5.1 for control chart factors.)

$$\overline{\overline{X}} = \frac{\Sigma \widetilde{X}}{k} = \frac{4095}{25} = 163.8 \quad \overline{R} = \frac{\Sigma R}{k} = \frac{145}{25} = 5.8$$

$$\text{UCL}_{\widetilde{X}} = \overline{\overline{X}} + \overline{\overline{A}}_2 \overline{R} = 163.8 + 1.187(5.8) = 170.7$$

$$\text{LCL}_{\widetilde{X}} = \overline{\overline{X}} - \overline{\overline{A}}_2 \overline{R} = 163.8 - 1.187(5.8) = 156.9$$

$$\text{UCL}_R = D_4 \overline{R} = 2.574(5.8) = 14.9$$

$$\text{LCL}_R = D_3 \overline{R} = 0(5.8) = 0$$

Chart analysis

The median and range charts both show a very good state of statistical control. This is about as good a state of statistical control as you will see. There are no points outside the control limits, and there are no unnatural patterns of variation. Based upon these 25 samples, the process is very stable, predictable, and consistent.

> QUESTION: Does having the process in control, which is the same as stable, predictable, and consistent, mean that all the product is good (in specification) and that the capability is good?
>
> ANSWER: No, having the process in control, which is synonymous with stability, predictability, and consistency, does not necessarily mean that all the product is in specification. Let me use an analogy: I have a friend who plays golf 10 or 15 times a year, and his golf scores are stable, consistent, and predictable. He is a consistently poor golfer. But I think I would describe his capability as a golfer as poor. On a regular basis he shoots triple bogey golf.

There are many manufacturing processes that are consistently bad. These kinds of processes would be classified as in control, but they are not capable of meeting the requirements. Try to sort out the two different topics of statistical control and capability. They are two completely different subjects. Both are very important. If someone is using a control chart, the strategy should be to first attain statistical control and then evaluate the capability of the process. If the capability is not acceptable, then a plan to reduce common-cause variation must be developed. Usually this plan will require much management action to improve the process.

Concerning this example, it does appear that this process is easily meeting the specifications. None of the 75 measurements are close at all to either of the specifications. The capability of this process appears to be good, from scanning the individual measurements at the top of the control chart form and comparing the measurements to the specification (Fig. 7.7).

> QUESTION: In this example, samples of 3 were taken. The median and ranges were calculated and plotted. The control limits were calculated for the medians and ranges. All the medians and ranges were within the control limits. If you use the original readings and the medians and ranges, will not all the points always be inside the control limits?
>
> ANSWER: No, what you are suggesting is not logical. The control limits for the medians and the ranges are the natural limits for a process *that is in a state of statistical control*. Most processes are not in a state of statistical control; in these situations, there will be a number of points outside the control limits.

Capability calculations

$$\hat{\sigma} = \frac{\overline{R}}{d_2} = \frac{5.8}{1.693} = 3.43$$

$$C_p = \frac{\text{tolerance}}{6\hat{\sigma}} = \frac{33}{20.6} = 1.60$$

$$Z_{\text{USL}} = \frac{\text{USL} - \widetilde{\overline{X}}}{\hat{\sigma}} = \frac{181.5 - 163.8}{3.43} = 5.16$$

$$Z_{\text{LSL}} = \frac{\widetilde{\overline{X}} - \text{LSL}}{\hat{\sigma}} = \frac{163.8 - 148.5}{3.43} = 4.46$$

$$C_{pk} = \frac{Z_{\min}}{3} = \frac{4.46}{3} = 1.49$$

Capability analysis

Since the process shows a very good state of statistical control, it is proper to determine the capability of the process. The low specification is 148.5, and the high specification is 181.5. The total tolerance width is 33 g.

The correct d_2 factor must be determined for a sample size of 3. The d_2 factor in this case is 1.693.

We estimate a standard deviation to be 3.43 g. The formula for capability C_p uses 6 standard deviations in the denominator. The C_p

value is 1.60. Another way of expressing this is as follows: 160 percent of the width of the 6σ distribution of the individual measurements will fit between the specification limits.

The next step is to calculate the C_{pk} value. This will tell if the process is centered adequately. Recall that the average median $\widetilde{\overline{X}}$ of the process was 163.8 g. The target is 165 g. This process is very well centered. If a process is perfectly centered at the nominal specification, then C_p and C_{pk} will be the same. The farther a process is off center, the more the C_{pk} value will drop down from what the C_p value is.

The two preliminary steps to determine C_{pk} require that we determine how many standard deviations it is from the center of the process to the lower specification Z_{LSL}, and how many standard deviations it is from the center of the process to the upper specification Z_{USL}. The Z_{LSL} is 5.16, and this means that the lower specification is 5.16 standard deviations for the center of the process. The Z_{USL} value is 4.46. This means that the upper specification is up 4.46 standard deviations from the center of the process. The last calculation will give us C_{pk}. The smaller of the two Z values is divided by 3, and that answer is C_{pk}. In this case C_{pk} is 1.49.

In summary, the process looks pretty good as far as capability is concerned. The process should be adjusted closer to the target of 165. This adjustment will improve the quality with the existing amount of variation in the process. The goal of 1.67 minimum for the capability of the process has fallen a little short. There should be continued effort to reduce the variation. I also think the workers can begin to focus on another opportunity for improving quality elsewhere. This quality characteristic should continue being monitored to verify that the process continues to be centered close to the target and that the variation in the process does not increase.

7.5 Case Study: Foundry

A foundry that makes iron castings is having chronic problems with internal scrap and rework, customer complaints, over-budget costs, and low productivity. One process parameter being monitored is the compactibility of the molding sand. The target is 30 percent compactibility. It is believed that if the compactibility stays between 20 and 40 percent, this aspect of the process will not cause problems with the product. The specifications will be considered 20 and 40 percent.

A decision was made to use a Shewhart control chart for averages and ranges and to record three measurements during a 24-h period (Fig. 7.8). The calculations for the control limits are shown here:

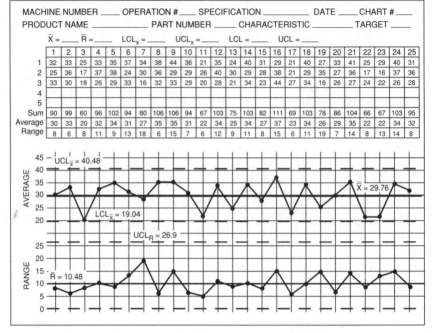

Figure 7.8 Average and range control chart for sand compactibility.

$$\overline{\overline{X}} = \frac{\sum \overline{X}}{k} = \frac{744}{25} = 29.76$$

$$\overline{R} = \frac{\sum R}{k} = \frac{262}{25} = 10.48$$

$$\text{UCL}_{\overline{X}} = \overline{\overline{X}} + A_2\overline{R} = 29.76 + 1.023(10.48) = 40.48$$

$$\text{LCL}_{\overline{X}} = \overline{\overline{X}} - A_2\overline{R} = 29.76 - 1.023(10.48) = 19.04$$

$$\text{UCL}_R = D_4\overline{R} = 2.574(10.48) = 27.0$$

$$\text{LCL}_R = D_3\overline{R} = 0(10.48) = 0$$

The average of this process is 29.76 percent. This tells us that the process is very well centered at the target of 30 percent. There is an unnatural observed variation. The process is not in a good state of statistical control. There should be more of the plotted averages closer to the centerline of the average chart. Recall that when a process is in statistical control, there should be about 68 percent of the averages in the central third portion of the average control chart. Unnatural variation is also observed with the average of sample numbers 3, 11, 22, and 23. It is not natural to see so many averages that close to the control limit. There should only be about 5 percent of the sample averages in the outer third zone of the control chart. Recall from earlier in the book that the distribution of averages will very closely approximate the shape of the normal distribution when the process is in statistical control.

Clearly the process is not consistently meeting specifications. There are nine individual measurements that are outside specifications of 20 and 40 percent. There are 25 samples of 3 for a total of 75 individual measurements, so there is an observed 12 percent out-of-specification rate. Notice that both control limits for the averages (40.48 and 19.04 percent) are outside the specifications. Also notice that the upper control limit for the ranges is greater than the specification width. These are indications that the process is producing nonconforming material.

Loss calculations

To develop a baseline for the loss due to this process, we arbitrarily use $1.00 as the k factor in the loss function. The average of the process, based upon the measurements from Fig. 7.8, is 29.8 percent, and the standard deviation is 6.32 percent.

The formula for NIB characteristics is

$$L = k[(\bar{y} - m)^2 + \sigma^2]$$

$\bar{y}$ = average m = target σ = standard deviation

$$L = \$1.00[(29.8 - 30.0)^2 + 6.32^2] = \$1.00(0.04 + 39.94) = \$39.98$$

The great majority of the loss is due to the large variation in the process; very little is due to the process being off target.

Figure 7.9 shows the 75 individual measurements and the 25 plotted averages.

It would not be proper to say that the process is not capable because the process is not yet in a good state of statistical control. It would be correct to say that the process is out of control (unstable) and is producing some out-of-specification product. Since the process is not in a good state of statistical control, it would not be proper to calculate capability for the process at this time. If we follow the standard formulas and do the mathematics properly, we are led to believe that the percentage nonconforming is better than the observed percentage.

In the next 12 months, some changes to the process were made that the team thought from their years of experience might reduce the process variation observed. The next control chart we see (Fig. 7.10)

174 Chapter Seven

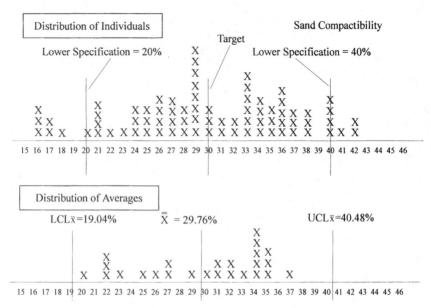

Figure 7.9 Distribution of individual measurements compared to specification limits and distribution of 25 averages compared to the calculated control limits.

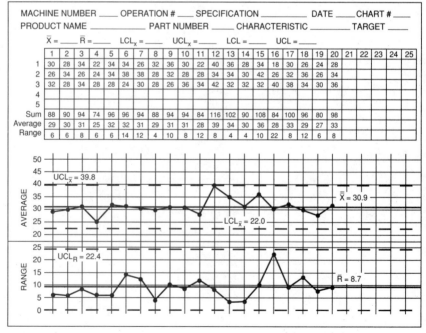

Figure 7.10 Average and range control chart for sand compactibility after process improvements.

is from the same process about a year later. The process showed signs of improvement, so it was appropriate to recalculate new tighter control limits for the improved process.

When we compare the control limits from the earlier control chart for sand compactibility, we notice the following:

1. The process has shifted slightly higher. The center of the process now is 30.9 percent. It was 29.76 percent.
2. The variation of the sand compactibility has become smaller. The average range was 10.48 percent. The average range for the process is currently 8.7 percent. Since the average range has become smaller, we use this value in all the calculations for the control limits. The new control limits for this improved process will be tighter than the limits for the first control chart.

$$\bar{\bar{X}} = \frac{\Sigma \bar{X}}{k} = \frac{618}{20} = 30.9$$

$$\bar{R} = \frac{\Sigma R}{k} = \frac{174}{20} = 8.7$$

$$\text{UCL}_{\bar{X}} = \bar{\bar{X}} + A_2 \bar{R} = 30.9 + 1.023(8.7) = 39.8$$

$$\text{LCL}_{\bar{X}} = \bar{\bar{X}} - A_2 \bar{R} = 30.9 - 1.023(8.7) = 22.0$$

$$\text{UCL}_R = D_4 \bar{R} = 2.574(8.7) = 22.4$$

$$\text{LCL}_R = D_3 \bar{R} = 0(8.7) = 0$$

3. There are no averages or ranges outside the calculated control limits. The only unnatural variation observed in the control chart might be that the averages for the first half of the control chart seem to "hug" the centerline.

The current process shows a better state of statistical control than the previous chart showed. There is a stronger basis for now evaluating the process capability.

Figure 7.11 shows the distributions of both averages and individuals. It is obvious that the variation in the process has been reduced, and the process is closer to a state of statistical control. The distribution of individuals is somewhat bimodal. The other interesting observation is that all the individual readings are even numbers. This should be investigated as to the cause of this unnatural condition.

$$\hat{\sigma} = \frac{\bar{R}}{d_2} = \frac{8.7}{1.693} = 5.14$$

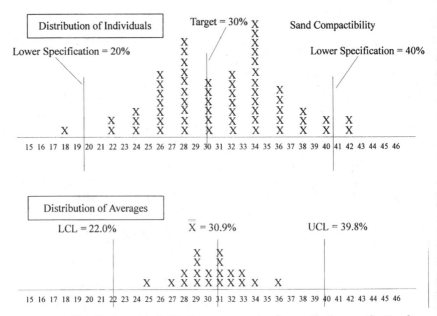

Figure 7.11 Distribution of individual measurements along with the specification limits, distribution of sample averages, and the new tighter control limits for the improved process.

$$C_p = \frac{\text{tolerance}}{6\hat{\sigma}} = \frac{20}{30.84} = 0.65$$

$$Z_{\text{USL}} = \frac{\text{USL} - \overline{\overline{X}}}{\hat{\sigma}} = \frac{40 - 30.9}{5.14} = 1.77$$

$$Z_{\text{LSL}} = \frac{\overline{\overline{X}} - \text{LSL}}{\hat{\sigma}} = \frac{30.9 - 20}{5.15} = 2.12$$

$$C_{pk} = \frac{Z_{\min}}{3} = \frac{1.77}{3} = 0.59$$

The C_p value is now 0.65. This means that 65 percent of the width of the distribution of individuals will fit inside the specification limits. It does not mean that 65 percent of the product is within specifications. To determine the percentage of product outside the specification, the Z values for the upper and lower specifications must be calculated. Then we look up those values in the Z table to determine the percentage outside the specification limits. Now Z_{USL} is 1.77. This means that the upper specification limit is up 1.77 standard deviations from the center of the process. And Z_{LSL} is 2.12, which means that the upper specification is 2.12 standard deviations from the center of the process.

Next we should look up the Z values to determine the estimated proportion of product outside the specifications. There is 3.84 percent of product outside the upper specification of 40 percent, and there is 1.74 percent of the product that is outside the lower specification. The total estimated proportion of product outside the specifications is 5.58 percent. The actual observed proportion out of specification is 3 of the 60 individual measurements, which is 5.0 percent. The gap between the theoretical and the reality is closing, which usually indicates that the process is getting closer to a state of perfect statistical control.

The proportion of out-of-specification product has come down due to less variation in the process. Time should be taken to calculate the loss for the process now that the process shows a state of statistical control. We will use a k factor of $1.00, just as we did in the earlier calculation for the loss.

$$L = k[(y - m)^2 + \sigma^2]$$

$$= \$1.00[(30.9 - 30.0)^2 + 5.14^2] = \$1.00(0.81 + 26.52) = \$27.33$$

The loss originally was $39.98, and it is currently $27.33, which is more than a 30 percent reduction in loss. This process is considered unsatisfactory, but progress is being made on the journey of continuous process improvement.

> QUESTION: I am confused about one thing concerning this example. I thought the control limits were 3 standard deviations from the centerline. On the last average chart, the upper control limit for the averages is 39.8 percent, and the centerline is 30.9 percent. The distance from the centerline of the control chart to the upper control limit is 8.9 percent. If that distance is 3 standard deviations, then 1 standard deviation would be 2.97 percent. This is where the point of confusion occurs. During the calculations for capability, we just determined that the standard deviation is 5.14 percent. These two numbers are not even close to each other (2.97 versus 5.14 percent). Why is there a difference, and which is the correct value for the standard deviation?
>
> ANSWER: The problem is that we are combining two completely different topics: control and capability. Concerning the control chart, it is true that the control limits are 3 standard deviations from the centerline of the chart. The important point is that the distance is 3 standard deviations of the averages, not individuals. The data points on the average chart are averages, so it is logical that the limits are 3 standard deviations of the averages from the centerline.

When the topic of control has been completed and the subject changes to capability, we are concerned with the extent to which the distributions of individual readings are meeting the specifications. Your determination that a standard deviation for the average chart is

2.97 percent is correct. But that is a standard deviation for the averages. The calculation for standard deviation during the capability phase was correct (5.14 percent), but this is a standard deviation for the individual readings. The central limit theorem tells us that there is a relationship between the standard deviation of the averages and the standard deviation of the individuals.

$$\sigma_x = \sigma_{\bar{x}} \times \sqrt{n} \quad \text{or conversely} \quad \sigma_{\bar{x}} = \frac{\sigma_x}{\sqrt{n}}$$

The sample size on the control chart is 3. The square root of 3 is 1.732. If we multiply the standard deviation of the averages by the square root of the sample size, we should get an answer very close to the value for the standard deviations of the individuals: 2.97 × 1.732 = 5.14 percent. There is no conflict between these two values for standard deviation. One is the standard deviation for the averages. The other is the standard deviation of the individuals. The person evaluating the process must be very careful to differentiate between statistical control and capability.

The process improvement effort to improve the quality of castings and reduce scrap and rework was successful to a point. At this stage the team seemed to hit a brick wall and did not know what to do to improve the processes. The internal constraints within the company did not allow for purchase of new capital equipment. It was almost a year later when the team conducted a multifactor designed experiment. Large gains in reducing the variation were attained. The average of the compactibility after the experiment was 31.67 percent, and the standard deviation was 2.78 percent. The capability of the process after the design experiment was $C_p = 1.20$ and $C_{pk} = 1.05$. This process now is barely capable. The loss should be recalculated now.

$$L = k[(\bar{y} - m)^2 + \sigma^2]$$

$$= \$1.00[(31.67 - 30.0)^2 + 2.78^2] = \$1.00(2.79 + 7.73) = \$10.52$$

Summary

In the foundry example, the objective was to improve the quality of the product, and the logical strategy is to improve the process that yields the castings. The compactibility of the sand is a process parameter, not a product quality characteristic. During this effort, many process parameters were concentrated on centering the process around the target and reducing variation. During the initial stages of charting the sand compactibility, the process would be considered very bad; later in the campaign to improve quality, the process would be considered poor. After the designed experiment the process would be considered barely capable. There has been much written about continuous process improvement. Improving a process from terrible to poor to barely capable is what continuous process improvement is all about. This is not to say that the journey is complete. This foundry was at a 20 percent internal scrap and rework rate when the journey began. The team was able to bring the scrap and rework rate down to slightly over 6 percent. During this time external customer satisfaction improved, and the cost of the product was reduced.

The loss due to the compactibility of the sand was originally calculated with a baseline value of $39.98. After the process was brought into a state of statistical control and process optimization was conducted by using the design of experiment, the loss had been reduced to $10.52. The process is still far short of a goal of achieving capability values of 1.67 minimum, but much improvement has taken place.

For purposes of explanation, the calculations evaluating the capability for this process will be carried out using the data from Fig. 7.8. Because the process is showing unnatural variation, we will think the process is producing a higher percentage of in-specification product than it really is. The capability values will be optimistically misleading since the process is not in control.

The C_p value addresses the question, Will the distribution of individual readings fit between the specification limits? Here the C_p calculated is 0.54. So we could state that 54 percent of the width of the 6σ distribution will fit between the specifications. It does not mean that 54 percent of the product is within the specifications.

$$\hat{\sigma} = \frac{\bar{R}}{d_2} = \frac{10.48}{1.693} = 6.19$$

$$C_p = \frac{\text{tolerance}}{6\hat{\sigma}} = \frac{20}{37.14} = 0.54$$

$$Z_{USL} = \frac{USL - \bar{\bar{X}}}{\hat{\sigma}} = \frac{40 - 29.76}{6.19} = 1.65$$

$$Z_{LSL} = \frac{\bar{\bar{X}} - LSL}{\hat{\sigma}} = \frac{29.76 - 20}{6.19} = 1.58$$

$$C_{pk} = \frac{Z_{min}}{3} = \frac{1.58}{3} = 0.53$$

When we calculate Z_{USL}, we get 1.65. This means that the upper specification limit is 1.65 standard deviations from the center of the process. The next step is to look up the proportion out of the upper specification. The Z table tells us to expect 4.94 percent of the product out of specification. Then Z_{LSL} is 1.58, and the estimated proportion out of specification is 5.71 percent. The estimated total percentage out of specification would be 4.94 + 5.71 = 10.6 percent. The actual percentage out of specification was 12 percent, so there is a minor discrepancy between the actual observation and the theoretical estimate. The reasons for this conflict could be these:

1. The process is not in a state of statistical control.
2. Since the sample size was 3, possibly the distribution of averages was not transformed to the normal distribution. A sample size of 4 or more would have more strongly guaranteed the transformation of the averages to the normal distribution.
3. Mistakes were made in the plotting of the data or in the calculations.

The analysis of this control chart (Fig. 7.8) goes something like this:

1. The averages for the process are showing unnatural variation. There are fewer than expected averages in the middle third of the average chart. The averages of samples 3, 11, 22, and 23 are all in the outer third of the control chart. Except for these two things, the process seems in a fairly good state of statistical control.
2. The process is well centered on the target. The average is 29.76 percent, and the target is 30 percent. [Notice that the C_p value (0.54) and the C_{pk} value (0.53) are quite close to the same number.] This indicates that the process is very well centered.
3. The process is producing a high proportion of nonconforming product (12 percent). This process would be considered unsatisfactory.

There must be an effort to reduce process variation. The team assigned to this project must identify factors in the process and decide on an action plan that will result in reduced variation of the compactibility of the molding sand.

7.6 Multivary Charting

Multivary charting analysis is a graphical charting method that is very efficient at minimizing the variation within a product. Traditional Shewhart control charts do an excellent job of evaluating the variation from product to product or from time to time. Causes of variation can be grouped into one of several stratified families of variation.

1. *Within-piece variation (positional).* An example of this family of variation would be eccentricity, flatness, thickness, or taper. This type of characteristic occurs where the magnitude of variation changes within the part. Examples of this type of characteristic are the taper, flatness, eccentricity, parallelism, and squareness. The target for this

type of characteristic is zero. Actually the same characteristic is being measured in many places within the product. If this is the situation, then we are trying to minimize the unwanted within-piece variation. Hundreds of companies today are trying to use Shewhart control charts for variables for these types of characteristics when in reality a more appropriate statistical tool is a multivary chart or possibly even a simple check sheet.

2. *Piece-to-piece variation (cyclical)*. This family of variation is observed among pieces produced during a short time. An example is the length of 10 pieces of cloth produced consecutively. If we are concerned about batch-to-batch or lot-to-lot variation, then we most likely want to investigate *cyclical variation*.

The equivalent of measuring piece-to-piece variation when using the Shewhart control chart is the range. The range is an indicator of variation within the sample.

3. *Time-to-time variation (temporal)*. This type of variation is related to the difference observed in products produced at different times of the day or on different days. Product made on different days will vary due to changes in operations or materials or environmental factors.

The equivalent to time-to-time variation when one is using a Shewhart control chart is the differences from average to average.

4. *Process stream to process stream*. This type of variation occurs when two or more different process streams are producing the same product (multiple fixtures, spindles, machines, cavities, and stations). This is another situation where many companies are trying to use Shewhart control charts, and the control charts are not designed to control manufacturing processes with multiple streams. The result is meaningless information when we should be really using a multivary chart or possibly a check sheet.

In this section, you will learn how to construct a multivary chart and then analyze the pattern of variation so that the factors causing this unwanted variation in product can be reduced or eliminated.

Purpose

A multivary chart analysis will identify which type of variation is having an adverse effect on the products and processes. This in turn will quickly reduce the larger number of process variables to a few prime suspects which can be investigated further. A multivary analysis allows you to break down and evaluate the many sources of variation that contribute to the overall variability in the process. You accomplish all this by constructing a graphical display of the process variation. The next step is to examine the graphical display to determine whether nonrandom patterns of variation exist. Nonrandom patterns of variation should be eliminated or at least minimized.

Sampling for multivary charting

The following information should be thought of only as suggested guidelines. The suggestions are not cast in concrete, but time has proved that they are a good baseline to start from.

1. Sample size should be 3, 4, or 5. It is suggested that the products be consecutively produced.
2. The time between samples should be long enough that variation is observed from sample to sample.
3. A minimum of 15 measurements should be taken during the evaluation of the process.

Visual analysis

Visual analysis of multivary charting is rather simple if you remember the goal while using the multivary charting method:

1. Minimize within-piece variation.
2. Minimize variation from product to product.
3. Minimize variation from sample to sample.
4. Minimize variation from process stream to process stream.

Example 7.1: Crankshaft bearing eccentricity. An engine manufacturer is suffering with high scrap, rework, and field failures due to crankshaft failures, due to an eccentricity (out-of-round) problem with the main bearing journals. All the measurements are taken from machine 2183. The measurements are in millionths of an inch (0.000001 in); the eccentricity is measured in the quality assurance laboratory. The sampling procedure is to measure and record the eccentricity readings on all seven main bearing journals once per hour.

Note: Main bearing 1 is the front end of the crankshaft, and bearing 7 is at the back end of the crankshaft.

The eccentricity readings of six crankshafts are shown below.

Part no.	Bearing number						
	1	2	3	4	5	6	7
2	80	140	180	200	140	100	100
3	60	100	120	140	140	100	40
10	100	120	120	180	160	120	80
11	100	160	180	180	120	120	80
18	80	80	140	180	160	140	100
19	40	60	100	140	100	80	60

Remember that the measurements are in millionths of an inch (for example, 120 = 0.00012 in). The measurements are plotted on the multivary chart (Fig. 7.12).

Variable Control Charts, Multivary Charting, and Precontrol Charts 183

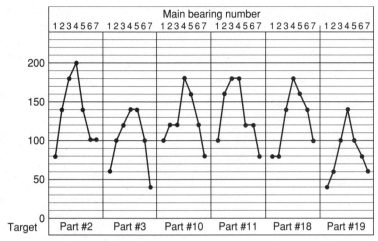

Readings are in millionths of an inch (150 =.00015″)
Specification .00015″ max. (150)
Target = 0

Figure 7.12 Multivary chart showing where the eccentricity in the crankshaft is (by bearing number).

Multivary analysis

The target for this type of quality characteristic (eccentricity) is zero. This is a smaller-is-better type of characteristic. The repetitive nonrandom pattern of variation is that of within-piece variation. Main bearing numbers 3, 4, and 5 are consistently the bearings with the most eccentricity.

We must further refine the tracking down of the cause of excessive eccentricity. The members of the process improvement team decided that the next level of analysis should be to gather and analyze data to determine whether there is any nonrandom pattern of variation by looking at eccentricity radially.

Figure 7.13 shows the eccentricity of main bearing number 4 only. The amount of eccentricity is recorded every 60°. *Note:* The centerline of the oil pump keyway is 0°.

Part no.	Radial location					
	0°	60°	120°	180°	240°	300°
1	120	140	200	160	120	120
2	80	100	160	140	120	60
7	40	80	140	120	100	80
8	20	60	120	100	60	40
13	80	100	120	100	80	60
14	40	80	140	120	60	40
19	120	160	200	160	120	100
20	80	120	160	160	100	40

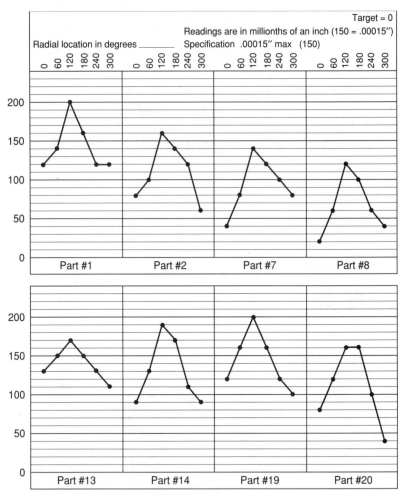

Figure 7.13 Multivary chart showing where the eccentricity is located (radial location).

Figure 7.13 shows the eccentricity of the number 4 main bearing from eight different crankshafts.

Chart analysis

A nonrandom pattern of variation is observed. The eccentricity is greatest at 120° from the keyway of the oil pump. Brainstorming sessions were held. One suggestion was that an out-of-balance condition in the crankshaft forging was causing the "whipping" action of the forging as it rotated in the grinding machine. There was a need to rework the forging dies to reduce the out-of-balance condition of the rough crankshaft forging. The eccentricity of the ground crankshaft was reduced by more than 40 percent with this improvement in the process. The multivary chart helped greatly to isolate the cause of excessive within-piece variation.

Summary

The multivary charting technique is a simple yet powerful aid in the reduction of variation. The objectives for use of the multivary charting method are different from those for Shewhart control charts. Determine which family of variation is causing the excessive unwanted variation:

Within-piece variation

Piece-to-piece variation

Time-to-time variation

Variation from process stream to process stream.

Once the family of variation has been determined, the task of finding the cause is greatly simplified.

7.7 Precontrol

Precontrol is another simple charting technique that can be used when measurements are variable, not attribute. The precontrol technique does not use calculated control limits, unlike the Shewhart control chart method. The limits are referred to as *precontrol limits*. These precontrol limits are for individual measurements. Individual measurements are plotted when the precontrol charting method is used. This is different from the Shewhart control charting method, which typically charts averages and ranges. The precontrol limits for individuals are located halfway between the midpoint of the specification and the specification limits. This portion of the chart is often colored green. The green portion is the "go" portion of the chart. The yellow (caution) portions of the chart are the outer 25 percent of the specifications. The red zone (stop) is for out-of-specification readings.

Example 7.2 The specification for the weight of a molded part is 260 ± 6 g. The low specification is 254 g, and the high specification is 266 g. In this example, we will assume that:

1. The distribution of individual part weights is the general shape of the normal distribution.

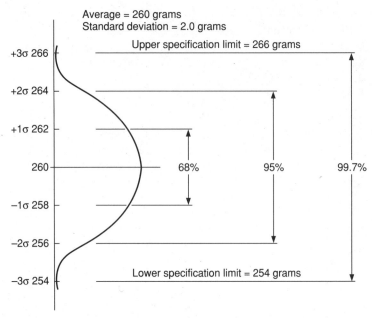

Figure 7.14 Distribution of part weight with theoretical area under the normal curve.

2. The distribution is very well centered at the midpoint of the specification.

The process distribution is the same width as the specification. In other words, the specification limits of 254 and 266 g are exactly 3 standard deviations from the center of the process. Thus the average is 260, and the standard deviation is 2 g.

Recall from earlier in the book that the area under the normal distribution, shown in Fig. 7.14, is as follows:

$$\overline{X} \pm 1\sigma = 68\% \qquad \overline{X} \pm 2\sigma = 95\% \qquad \overline{X} \pm 3\sigma = 99.7\%$$

The precontrol limits will be the middle 50 percent of the tolerance band (the green zone). The lower precontrol limits will be 257 g. The upper warning limits will be 263 g.

Figure 7.15 shows the location of the precontrol limits in relation to the weight distribution of the molded parts. Here, the precontrol limits of 263 and 257 g happen to be 1.5σ away from the center of the process (260 g), which also happens to be the target. The area under the normal distribution that is between +1.5σ and −1.5σ is 86.6 percent. The portion of the normal distribution that is above the upper warning limit is 6.7 percent. Likewise, the area of the normal distribution that is below the lower warning limit is 6.7 percent.

The usual procedure for using the precontrol charting technique has two phases: qualification and sampling.

qualification When the process is set up and running satisfactorily, measure five consecutively produced products. All five measurements must be within the middle 50 percent of the tolerance band (the green zone) to satisfy the qualification phase. The process average is 260g. The standard deviation is 2 g, so approximately 87 percent of the measurements are within the precontrol limits.

Variable Control Charts, Multivary Charting, and Precontrol Charts

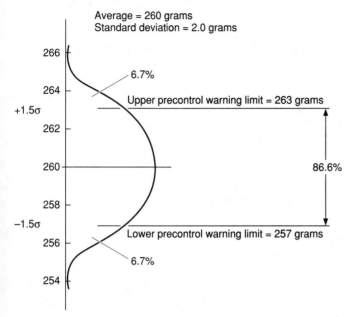

Figure 7.15 Distribution of part weight with precontrol warning limits drawn in.

Given this, the probability of passing the qualification phase is $0.87^5 = 0.49$. There is a little bit less than a 50 percent chance of passing the qualification phase. If all five measurements are not within the precontrol limits, the setup of the process should be reviewed and any adjustment made, if needed. Then the qualification phase should be repeated.

If all five measurements are within the precontrol limits, the sampling phase can begin. Periodic samples will be measured. Take two consecutively produced products. If the first measurement is within the limits, the second measurement is not taken. Figure 7.15 shows that the probability of a reading being outside the precontrol limits is about 13.4 percent. If the first sample is outside the precontrol limits, measure and plot the second reading. If the second measurement is within the limits, continue production until the time for the next sample. If the second measurement will also be outside the precontrol limits, corrective action must be taken. The probability of two measurements coming from a process with an average of 260 g and a standard deviation of 2 g being outside the precontrol limits is quite small: $0.134^2 = 0.018$. There is less than a 2 percent chance of two measurements being outside the precontrol limits.

sampling The A portion of Fig. 7.16 shows that the qualification phase has been satisfied by five consecutive products in the green zone. The sampling can now begin. In portion B to D of Fig. 7.16, the first measurement is in the green zone. It is not required to measure or plot the second reading. In portion E of Fig. 7.16 the first measurement in the sample is in the yellow (caution) zone. The second measurement must be taken and recorded. The second reading is in the green zone. The sampling can continue. In portion F, the first measurement is in the green zone. In portion G the first reading of the sample is in the yellow

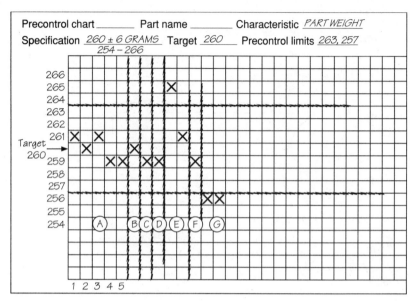

Figure 7.16 Completed precontrol chart.

(caution) zone. The second measurement must be taken. The second reading is also in the yellow zone. Corrective action must be taken on the process. Anytime an out-of-specification measurement is observed, supervisors should be informed so that corrective action can be taken.

If the process worsened by

1. An increase in variation
2. The process drifts off target

then the probability of meeting the requirements of the qualification phase and the sampling phase would go down. Precontrol should not be used when the 6σ distribution of individuals does not fit within the specification.

If a process has good capability, there should be very little problem meeting the qualification and the sampling phase.

The main advantage of precontrol is that it is a very simple technique for everyone to use. Some of the disadvantages are that it

1. Uses specification limits (middle 50 percent) to determine when to take action on the process
2. Does not identify special-cause variation
3. Does not drive a company toward continuous improvement
4. Should not be used on processes that are producing nonconforming product

7.8 Summary

1. When a control chart for variables is used, the *objective* is to
   a. Improve the process so that it is in true statistical control.
   b. Adjust the process close to the customer's target.
   c. Improve the process to the extent that common-cause variation is small enough that the process is considered satisfactory.
2. Many world-class companies feel very good going beyond meeting blueprint specifications. These companies use the calculated control limits to determine when investigation and corrective action are necessary.
3. A good customer complaint report, scrap report, rework report, or assembly line problem feedback report can give guidance as to where to use a variables control chart.
4. If a process is close to the target and in control, the control limits should be left alone unless we see process improvement (reduction in the sample ranges, or averages getting even closer to the target).
5. Do not confuse statistical control with a satisfactory process. Synonyms of statistical control are *stable, predictable,* and *consistent.*
6. Do not get locked into thinking that there is only one type of variables chart, sample size, or sampling scheme that is correct for all processes.
7. It is very important that information about the process be noted on the control chart. It is just as important that someone analyze the reason for out-of-control conditions to be summarized and prioritized and correction action to be initiated to prevent them from happening in the future. This is how processes are improved!
8. Ensure that the operator, supervisor, and all other key people know how to visually analyze the control chart for nonrandom patterns of variation so that investigation and corrective action can be undertaken.
9. Use the formulas described in this chapter. If for some reason you choose to use some other method of calculations, do not refer to the lines on the control chart as control limits, nor should you refer to the chart as a control chart.
10. Whenever the objective is to reduce within-piece variation, strongly consider using a multivary chart to graphically determine the pattern of variation within the part.
11. A precontrol chart should be considered if there is a need just to monitor the process, not necessarily improve it. Only use precontrol charts on processes or quality characteristics that can easily meet the specifications.

Chapter 8

Attribute Control Charts

In gathering data at the beginning stages of a company's process improvement effort, an attribute control chart is a logical statistical tool to use. If a company does not have any information that can direct the process improvement efforts to specific departments, machines, processes, or quality characteristics, then attribute control charts in conjunction with a check sheet categorizing the reason for nonconformance are one of the best tools to use. Attribute control charts have strong applications in nonmanufacturing situations (service industries and administrative situations).

In recent years there is a hesitancy to use the term *defective* or *defects* in evaluating a product, process, design, or service performed. Much of this hesitancy is due to the complicated issue relating to product liability. The terms *nonconforming* and *nonconformities* are used rather than defective or defects.

8.1 How Attribute Control Charts Fit into the Process Improvement Effort

Attribute information is typically gathered when a product, process, or service is classified as either acceptable or not acceptable. There are many situations in a manufacturing environment using attribute-type gaging (go/no-go gages); therefore it would not be possible to use variable-type control charts.

Attribute-type control charts can be easily used when a visual inspection is performed. Attribute control charts can be used to audit the level of quality of almost any administrative process: purchase orders completed, work instructions written, engineering drawings completed, attendance levels, quotations submitted, on-time delivery performance. This type of information is not as powerful as variable readings. The reason is that when measuring attribute-type information, we are "detecting" the frequency of things that have gone bad. The goal is to prevent all bad things from happening, in other words, to improve the quality level until zero nonconformities are achieved. The goal of zero nonconformities is met by preventing more and more of the nonconformities. Control charts for attributes do not really give us a prevention system. The main power of the attribute control chart information is that these charts tell us both whether the quality level is stable and consistent and what the average quality level is. Then the company can prioritize the opportunities for improvement. Usually when we go to work on the "big opportunities" that the attribute charts have highlighted, it will be necessary to go back into the process and gather variable-type information or other in-depth data by using additional quality tools (scatter plots, check sheets, analysis of customer surveys or employee surveys, Pareto analysis).

One very important point that is sometimes overlooked in the use of attribute control charts is the need to record additional detailed information as to the cause of nonconformance. This is a requirement so that action can be taken to improve the process.

8.2 Types of Attribute Control Charts

Many times the question is posed: What type of attribute control chart should I use? There is no single answer. But there are certain constraints that must be followed which may reduce the list of attribute-type charts that are appropriate.

The p chart is used when we want to chart the proportion of nonconforming items. The sample size from subgroup to subgroup can vary. The formulas for the control limits are based upon the binomial probability distribution.

The np chart is used when the number of nonconforming items will be charted. The sample size must remain constant. The np chart is a little bit simpler to understand because the number of nonconformities is recorded and plotted directly on the chart.

The c chart is used when the number of nonconformities will be charted. The sample size must stay constant. The control limits for the c chart are not based upon the binomial probability distribution, but the Poisson probability distribution.

The u chart is used when the number of nonconformities will be charted. The sample size can vary from sample to sample.

8.3 Considerations Prior to Use of Attribute Control Charts

The p chart for proportion nonconforming

A p chart is where we record and evaluate the *proportion* of nonconforming items.

First, decide on what product line, characteristic, or process will be monitored.

Second, ensure that the people gathering the data are consistent in their evaluation concerning how to tell conforming from nonconforming items.

Third, determine the sample size. If possible, keep the sample size consistent throughout the duration of the process evaluation. On the chart we must record the sample size n, the number of products rejected np, and finally the proportion rejected p:

$$p = \frac{np}{n}$$

When an attribute control chart is used, rather large samples must be taken. The sample size must be considered along with the proportion nonconforming. A good rule of thumb is that the average number of nonconforming $n\bar{p}$ should be at least 5. This will allow changes in the quality level to be observed on the control chart. The required sample size can be estimated if we know the average level of quality $\bar{p}$.

Example 8.1 A sewing process that has a daily production rate of 30 units per day is running at about a 0.04 (4 percent) level of nonconformance. What should the sample size be?

$$\text{Sample size} = \frac{n\bar{p} \text{ required}}{\bar{p}} = \frac{5}{0.04} = 125$$

A sample size of 125 will be sufficient. This means that once every 4 days a point could be plotted on the control chart.

When an attribute-type control chart is used, it is okay to perform 100 percent inspection of the product. Odds are that the production rates will vary from day to day. This situation is likely in both a manufacturing and an administrative circumstance.

> QUESTION: Does this mean that 25 samples must be gathered before the control limits can be calculated? This would require 100 days of production because of the low production rate. This does not sound like a good decision to wait this long to determine whether the process is in control.

ANSWER: Waiting 100 days before action is taken to improve a process is not being suggested. While the data are being collected for the control chart, it is assumed that the check sheet portion of the control chart is being completed and that the cause of nonconforming items is being recorded. The analysis of the check sheet portion can take place after 5 or 10 subgroups have been recorded on the control chart. If the information on the control chart or check sheet can lead to corrective action needed by the process, then by all means take the corrective action. Make some type of notation on the chart so that if there is an improvement in the level of quality, the cause of this improvement will be noted and can be taken into account in determining whether the process output is in statistical control. Remember, the true goal is to drive the process out of control on the low side so that the average quality level is closer to the target of zero.

There has been much discussion throughout this book about the normal curve. We should caution that in some situations, the distribution of the nonconforming proportion (p's) will not be the shape of the normal distribution. The guidelines for pattern analysis cannot be as strongly enforced.

Past experience has proved that when the average number of nonconformities np is 5 or greater, then the distribution will be the general shape of the normal curve. To ensure that this takes place, we might be required to increase the sample size. This relates to a phenomenon known as the *law of large numbers*.

The control limits are calculated so they are $\pm 3\sigma_p$ from the centerline of the chart. This type of chart will tell us whether the output of the process has a stable, consistent, and predictable quality output. But the really important decision that must be made by management is the following: *Is there much opportunity for improvement in the quality level based upon this attribute control chart? Is this one of our top quality and productivity problems?* If the answer is yes, then the management of the company must take action.

Once we have determined the vital few areas needing the greatest attention, we can use other statistical tools that will help lead us to where certain areas need more detailed examination, possibly by using flowcharts, histograms, run charts, control charts for variables, check sheets, scatter plots, or the design of experiment.

Fourth and last, *at least* 25 subgroups should be taken before the control limits are calculated. Also the period of time should be long enough to capture all the likely sources of variation affecting the process.

8.4 A *p* Chart with Constant Sample Size

A manufacturer of circuit boards has been detecting an unacceptably high number of circuit boards requiring manual rework of some soldered joints. All circuit boards are both functionally tested and visually inspected 100 percent. Every shift, 91 circuit boards are visually inspected, and the findings are recorded on the attribute control chart. The reason that the circuit board required touch-up of the wave soldered joints is recorded on the check sheet portion of the attribute control chart (Fig. 8.1). The calculations for the control limits are shown here:

Total number reworked:

$$\sum np = 207$$

Total number inspected:

$$\sum n = 2275$$

Average proportion nonconforming:

$$\bar{p} = \frac{\sum np}{\sum n} = \frac{207}{2275} = 0.091 = 9.1\%$$

$$\text{UCL}_p = \bar{p} + \frac{3\sqrt{p(1-p)}}{\sqrt{n}}$$

$$= 0.091 + \frac{3\sqrt{0.091(1-0.091)}}{\sqrt{91}}$$

$$= 0.091 + 0.0904 = 0.181 \quad (18.1\%)$$

$$\text{LCL}_p = \bar{p} - \frac{3\sqrt{p(1-p)}}{\sqrt{n}}$$

$$= 0.091 - \frac{3\sqrt{0.091(1-0.091)}}{\sqrt{91}}$$

$$= 0.091 - \frac{3(0.2876)}{9.54}$$

$$= 0.091 - 0.0904 \approx 0$$

Notice that in the denominator of the formula for the control limits the average sample size $\bar{n}$ is used. In this example, the sample size remains constant (91). So in this case either the sample size n or the average sample size $\bar{n}$ could be used. Recall that the p chart can be used if there is a constant sample size or if the sample size varies from subgroup to subgroup. The average sample size $\bar{n}$ must be used in the formulas for the control limits if the sample size does vary from subgroup to subgroup.

Chapter Eight

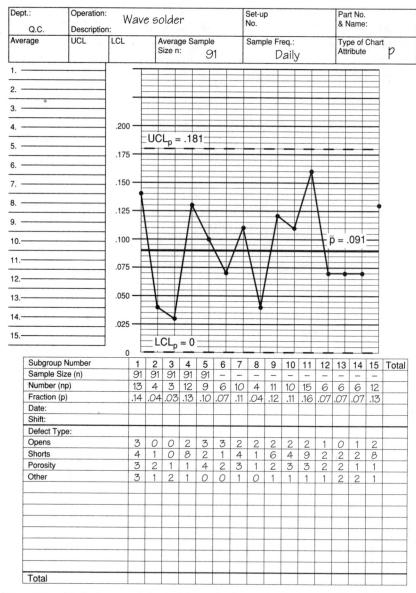

Figure 8.1a Attribute control chart with constant sample size (wave solder chart).

Attribute Control Charts

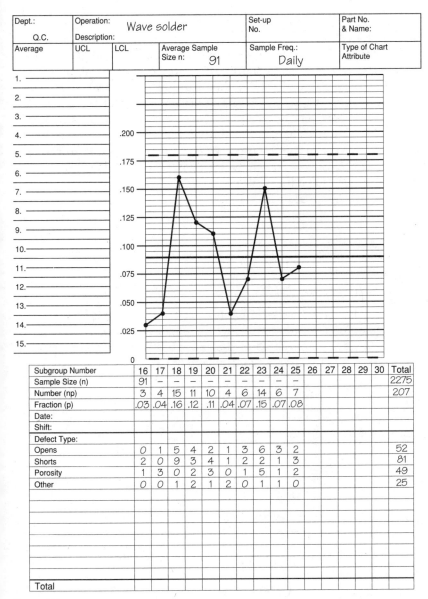

Figure 8.1b

Chart analysis

All the plotted points on the control chart are inside the calculated control limits. There are too few points in the middle third of the control chart. For a process to be in perfect statistical control, 68 percent of the points should be in the middle third of the control chart. There is 52 percent (13 of 25 points) in the middle third of the chart. The distribution of the data points from the chart is the shape of a skewed distribution, not the normal distribution.

This is the reason for the unexpected number of points in the middle third of the control chart. Shown below is a breakdown by cause of the circuit boards needing rework.

Cause	Frequency	Percentage	Cumulative percentage
Short circuits	81	39	39
Open circuits	52	25	64
Porosity	49	24	88
Other	25	12	100

A member of the process improvement team spent much time looking for some pattern hidden below the surface. Samples 1, 4, 11, 15, 18, and 23 were analyzed as to the cause of rework. The summary of samples 1, 4, 11, 15, 18, and 23 is shown here:

Subgroup number	Open circuits	Short circuits	Porosity	Other	Total
1	3	4	3	3	13
4	2	8	1	1	12
11	2	9	3	1	15
15	2	8	1	1	12
18	5	9	0	1	15
23	6	2	5	1	14
Total	20	40	13	8	81
Percentage	25	49	16	10	100

These six subgroups have an average rework rate of 15.4 percent. This is almost twice the average rework rate for all the subgroups. The main opportunity for improvement of short circuits becomes more obvious from this analysis. The manufacturer of the circuit boards is going to conduct a multifactor designed experiment to determine the factors contributing to short circuits as a cause of rework. The check sheet portion of the attribute control chart is vital if process improvement is to happen. The purpose of the attribute control chart is to tell whether the proportion of circuit boards is stable, consistent, and predictable. In other words, is it in statistical control?

Recall that the target for rework is zero. So there should be an effort to drive the process out of control on the *low side*.

QUESTION: When should the control limits be recalculated?

ANSWER: The control limits for the attribute control chart should be recalculated when the process has improved. There are two signs that the process has improved:

1. The average proportion nonconforming has come down closer to the ideal target of zero. Experienced practitioners suggest that a reduction in the average proportion nonconforming of 15 percent is enough to justify the recalculation of the control limits.
2. If there are points outside the control limits and the cause has been identified and corrective action has been initiated, then those points should not be considered in the calculations. Since these undesired high out-of-control points will not be used in the recalculations of the control limits, the new control limits will be lower. This is a good sign, because corrective action has been taken on the points that were out of control on the high side so they will not happen in the future. Therefore the quality level is improved, and it is logical that new control limits be calculated for the improved process.

8.5 A *p* Chart with Variable Sample Size

Figure 8.2 is a p chart for a nonmanufacturing situation. The process being evaluated is the timely payment of invoices. These delinquent invoices caused a financial penalty and adversely affected customer satisfaction. In all attribute-type charts, the target is zero. This is a smaller-is-better type of characteristic. Notice that at the bottom of Fig. 8.2 there is an area of the chart for keeping track of what department is responsible for the late payment. The sample size is the total number of invoices paid for the month. The sample size varies greatly from month to month. In this case, 100 percent inspection is being done.

Do as follows to work this problem:

1. Record the sample size for each month.
2. Sort the delinquent invoices by department at fault, and record that information in the check sheet portion of the attribute chart.
3. Calculate the proportion delinquent for each month.
4. Calculate the total delinquent invoices in the 25 samples:

$$\sum np = 221$$

5. Calculate the total number of invoices reviewed in the 25 samples:

$$\sum n = 15{,}671$$

200 Chapter Eight

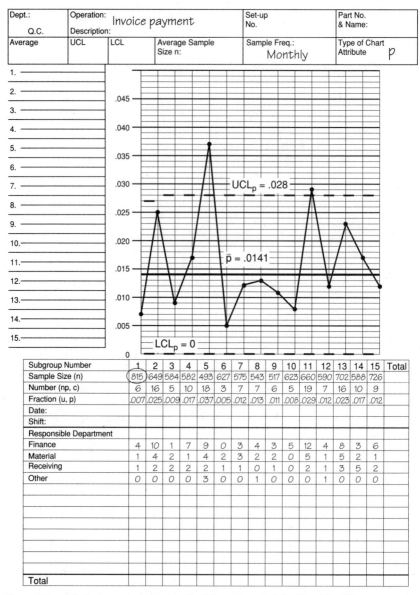

Figure 8.2a Attribute control chart with different control limits with different sample sizes (invoice payment example).

Attribute Control Charts

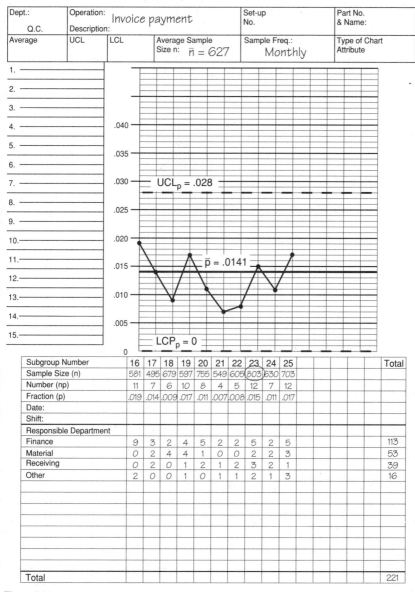

Figure 8.2b

6. Calculate the average sample size:

$$\bar{n} = \frac{\sum n}{k} = \frac{15{,}671}{25} = 627$$

7. Calculate the average proportion delinquent:

$$\bar{p} = \frac{\sum np}{\sum n} = \frac{221}{15{,}671} = 0.0141\,(1.41\%)$$

About 1.4 percent of the invoices on average are delinquent.

8. The control limits based upon the average sample size must be calculated.

$$\text{UCL}_p = \bar{p} + \frac{3\sqrt{\bar{p}(1-\bar{p})}}{\sqrt{\bar{n}}}$$

$$= 0.0141 + \frac{\sqrt{3 0.0141(1-0.0141)}}{\sqrt{627}}$$

$$= 0.0141 + 0.0140 = 0.0281 \quad (2.81\%)$$

$$\text{LCL}_p = \bar{p} - \frac{3\sqrt{\bar{p}(1-\bar{p})}}{\sqrt{\bar{n}}}$$

$$= 0.0141 - \frac{3\sqrt{0.0141(1-0.0141)}}{\sqrt{627}}$$

$$= 0.0141 - 0.0140 \approx 0$$

The control limits are computed with the average sample size of 627. In this example, the sample sizes vary from a high of 815 invoices to a low of 493 invoices. The control limits are dependent upon the sample size. If there are any sample sizes that are more than 25 percent different from the average sample size, then the control limits for these samples must be recalculated. If there are any samples greater than 784 or less than 470, control limits must be calculated for those samples. Samples 1 and 23 are greater than 788, so control limits for those 2 days must be calculated.

The second-to-last line of the calculations can be duplicated with only one change. The sample size for that day (815) must be used to determine the control limits. The control limits for the first sample with a size of 815 are

$$\text{UCL}_p = 0.0141 + \frac{0.35371}{\sqrt{815}} = 0.0265$$

$$\text{LCL}_p = 0.0141 - 0.0124 = 0.002\,(0.2\%)$$

The proportion of delinquent invoices for the first month (0.007) is within the control limits for that month. The limits for sample 23 are calculated in the same way as for the first sample. This calculation does not need to be done. The reason is that the sample size of sample 23 (803) is very close to the sample size of the first sample (815). The proportion delinquent for sample 23 is 0.015. Obviously this value will be inside the control limits. As the sample size increases, the control limits get closer to each other. Time can be spent on more important activities relating to improving this process.

Figure 8.2 shows the upper control limit of 0.0265 drawn in for the first sample. Many computerized statistical process control (SPC) programs will automatically calculate control limits for every sample size on the chart whenever the sample size changes.

Chart analysis

The process is considered out of control because samples 5 and 11 are above the upper control limit of 0.028. Also sample 2 is very close to the upper control limit. Of the 25 samples, 23 are within the calculated control limits. The mathematical theory states that when a process is in a perfect state of statistical control, 99.7 percent of the samples will be inside the control limits and there will be no unnatural patterns of variation. The process is close to being in a state of statistical control. It does seem that the level of delinquent invoices is becoming more stable during the last 14 samples. Unfortunately the proportion of delinquent invoices appears visually to be the same as in the first half of the chart. The last 14 points on the chart are hugging the centerline more closely. This means that there are fewer points close to the upper control limit. This is a good sign. Likewise, visually there are somewhat fewer points close to the lower control limit. This is a bad situation. Remember that the target for attribute control charts is zero.

The good thing about this type of chart is that in addition to plotting the proportion of delinquent invoices, we see a listing of the departments at fault. The check sheet portions of the attribute chart should be summarized and analyzed. Usually the vital few opportunities for improvement will be seen during the prioritizing of the information on the check sheet portions. If we were to prioritize the leading reasons for delinquent invoices, we would discover that the finance department was responsible for the great majority of these delinquencies. We should investigate the problem that the finance department is encountering so that the problem can be minimized in the future. The big opportunity for reducing the number of delinquent invoices is dependent upon improving the timeliness of processing the invoices in the finance department. Figure 8.3 is the Pareto chart showing the department at fault.

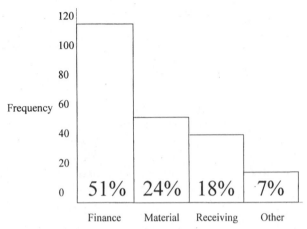

Figure 8.3 Pareto chart ranking department at fault (invoice payment example).

8.6 An *np* Chart with Constant Sample Size

The *np* chart is used to evaluate the number of nonconforming items in a subgroup. This chart is very similar to the *p* chart in which the proportion of nonconforming units is being charted. The only different requirement for the *np* chart is that the sample size must remain constant in all the subgroups. The *np* chart is used when we want to chart the number of nonconforming things. If we want to use this type of chart, the one restriction is that the sample size must remain constant.

Example 8.2 The circuit board example used in the explanation of the *p* chart will be used to explain how an *np* chart works. The average number of nonconforming items must be calculated.

$$n\bar{p} = \frac{\Sigma np}{k} = \frac{207}{25} + 8.28$$

where $n\bar{p}$ = average number nonconforming
Σnp = sum of nonconforming
k = number of samples

The formulas for the control limits for the *np* chart are as follows:

$$\text{UCL}_{np} = n\bar{p} + 3\sqrt{n\bar{p}(1 - \bar{p})}$$
$$= 8.28 + 3\sqrt{8.28(1 - 0.091)}$$
$$= 8.28 + 8.23 = 16.51$$
$$\text{LCL}_{np} = n\bar{p} - 3\sqrt{n\bar{p}(1 - \bar{p})}$$
$$= 8.28 - 8.23 = 0.05$$

Figure 8.4 is the *np* chart for the number of nonconforming occurrences. The calculated control limits have been drawn in. These control limits are 3 standard deviations away from the centerline of the control chart. Analysis of the *np* chart will be the same as it was when the data were analyzed earlier in this chapter by using the *p* chart (Fig. 8.1).

8.7 A *c* Chart for Nonconformities

A *c* chart is appropriate when there is a need to monitor the number of nonconformities in the sample. Now is a good time to explain the difference between nonconforming items and nonconformities.

A *nonconforming item* is one that is categorized as unacceptable. There may be one or more reasons why the item is classified as nonconforming. A *p* chart or an *np* chart is the type of attribute control chart to evaluate the proportion of nonconforming items or the number of nonconforming items.

Nonconformities are multiple occurrences, any of which would be the cause of classifying an item as nonconforming. The *c* chart and the *u* chart are the attribute charts that should be used to evaluate the magnitude of nonconformities.

It is quite possible that a nonconforming unit or item has more than one nonconformity.

In the two previous types of control charts explained (*p* chart and *np* chart), either the proportion of nonconformities or the number of nonconformities is charted. The sample size must remain constant when a *c* chart is used.

Example 8.3 A manufacturer of fiberglass truck camper tops has detected an increase in the number of complaints from dealers regarding cosmetic blemishes. Daily inspection log sheets from a year ago were analyzed, and they disclose that the average number of cosmetic blemishes per camper top was 4.9. This type of process evaluation will indicate the level of quality being shipped to the dealers. This effort is very much a detection activity.

A control chart was initiated at the end of the shipping department to determine whether the level of quality was in statistical control. The sample size is 1 camper top. The calculations for the control limits are shown here:

$$\bar{c} = \frac{\sum u}{k} = \frac{192}{25} = 7.68$$

where $\bar{c}$ = average number of nonconformities per sample
u = number of nonconformities
k = number of samples

$$\text{UCL}_c = \bar{c} + 3\sqrt{\bar{c}}$$
$$= 7.68 + 3\sqrt{7.68} = 16.00$$
$$\text{LCL}_c = \bar{c} - 3\sqrt{\bar{c}}$$
$$= 7.68 - 8.31 = -0.63 = 0$$

Chapter Eight

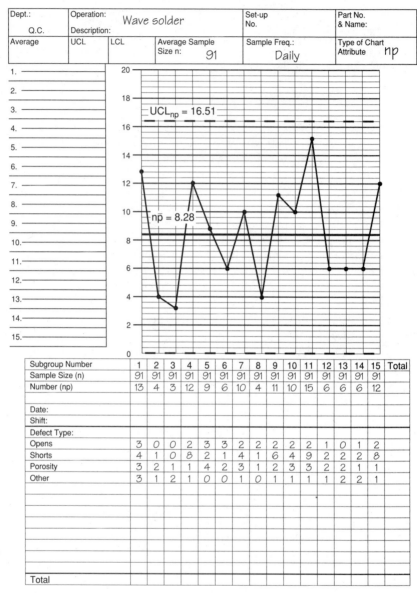

Figure 8.4a The np chart (sample size must remain constant) for the wave solder example.

Attribute Control Charts

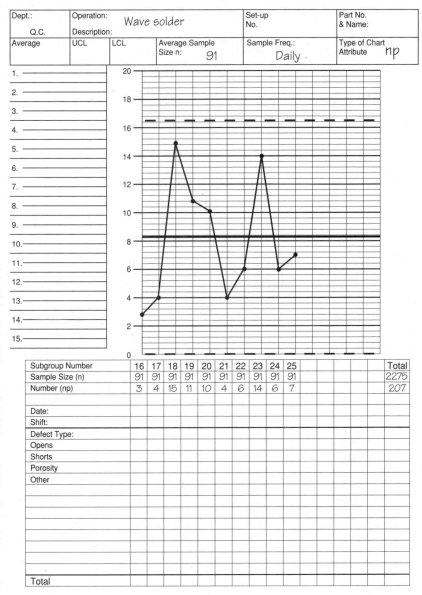

Figure 8.4b

There are no points outside of the control limits. Samples 4 and 10 are on the upper control limit. Other than that, the quality level of the camper tops does seem to be stable. The final inspector reported that of the 6 assembly dollies, dolly numbers 1 and 4 seem to always have a higher average number of cosmetic blemishes. The 25 measurements from the camper tops inspected on the control chart were sorted by assembly dolly number. The results are shown here:

Assembly dolly	Number of blemishes	Average number of blemishes
1	13, 10, 9, 9, 14	11.0
2	8, 3, 6, 3	5.0
3	7, 5, 9, 3	6.0
4	16, 16, 8, 12	13
5	2, 10, 5, 0	4.3
6	5, 7, 10, 2	6.0

The c chart (Fig. 8.5) highlights the assembly dolly number so that the readings can be traced back to each sample. Visually it is obvious that the measurements from dollies 1 and 4 are the process streams that contribute heavily to the high number of cosmetic blemishes for the product. The control chart would aid very little in process improvement without the detective work of identifying the sources of the blemishes.

Further investigation was done so that action could be taken to improve the level of quality of product coming from assembly dollies 1 and 4. The problems were worn parts causing jiggling and excessive movement on the assembly line and rough surfaces on the molds for assembly dollies 1 and 4. Repair work was done on the two dollies, and subsequent control charts indicated the average quality level had come down to 4.4 cosmetic blemishes per camper top.

QUESTION: Does this chart indicate that every camper top being shipped to the dealers has multiple blemishes? This seems to go against the principle of zero blemishes.

ANSWER: The first goal is to find out the average quality level and whether this quality level is in a state of statistical control. The c chart will do little by itself to improve the level of quality of the product. The primary purpose of the control chart is to determine whether special-cause variation is affecting the quality level. The control chart did flag the two readings right at the upper control limit (samples 4 and 10). Because all six of the process streams were grouped together on the control chart, the chart does not really do a good job of sorting out common-cause variation from special-cause variation. It is possible that just a simple tally sheet sorting the measurements by assembly dolly number would have pointed to the source of the opportunity of improvement without even using a control chart. If all the process streams exhibit the same level of quality and variation in quality level, then all the streams could be combined into one control chart. This is an example of a multistream process. Earlier in the book, we stressed that the process must be sorted by streams in the beginning stages of process evaluation. If the streams

Attribute Control Charts

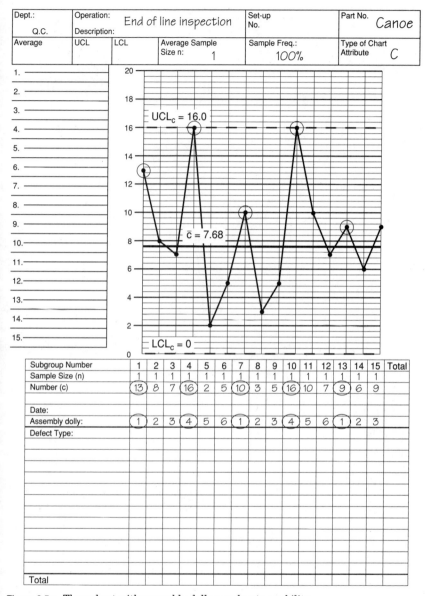

Figure 8.5a The c chart with assembly dolly number traceability.

Chapter Eight

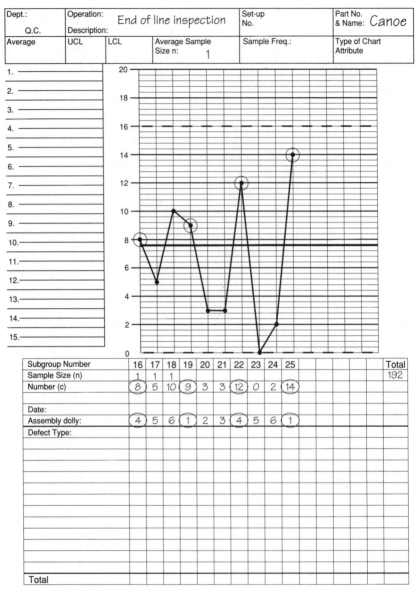

Figure 8.5b

are mixed, wrong conclusions can be drawn concerning both the statistical control and the capability of the process.

A c chart is used in a situation where we want to chart the *number* of nonconformities. One restriction on the use of this type of chart is that the sample size must remain constant.

A u chart is used when we want to chart the number of nonconformities *per unit*. The sample size can vary. This allowance for variable sample sizes is the only major difference between the c chart and the u chart.

8.8 Summary

An attribute control chart is a detection tool. When we use this type of control chart, we are recording the number or the frequency of bad things happening. The main objective when we use a control chart for attributes is to determine the average level of quality and whether that level is stable, consistent, and predictable, or, in other words, in statistical control.

We want to find out whether the variation in the quality level is caused only by common-cause variation (in control) or whether both common- and special-cause variation (out of control) affect the quality level. Once this knowledge is gained, an action plan can be developed to improve the process which will improve the level of quality.

One very important point is to attach a check sheet to the attribute chart that classifies the type of nonconformity, or the reason for rejection, so that the causes can be prioritized and corrective action can be initiated.

Remember that the target for attribute control charts is zero. Recall that the target for rework is zero. Record the reason that the product was classified as reject. Remember, the true goal is to drive the process out of control on the low side so that the average quality level is closer to the target of zero.

Chapter 9

Check Sheets

Check sheets are one of seven statistical tools to be used in a well-rounded quality improvement effort. Often check sheets are overlooked as too simple to be of significant benefit. Check sheets are an excellent tool to help in the stratification of data. Examples of stratification include

Multiple process streams

Long list of potential causes of rejections

Numerous product or process characteristics that require monitoring

Check sheets are an excellent tool to use at least in the early stages of the improvement process. A check sheet is a systematic method for collecting data. A check sheet must be simple, accurate, and well organized, and it must provide detailed information so that action can be taken on the process.

9.1 Constructing a Check Sheet

1. Write the goal or objective to be accomplished.
2. Identify the type of data you will collect. Will it be *variable* or *attribute* type? If variable data will be recorded, will the data be in line graph form or raw tabular form? If attribute measurements will be recorded, *determine exactly what you want to evaluate.*
3. Design the check sheet. (You must observe the process. A flowchart of the current process will highlight where data should be gathered and likely locations of the problems.) One of the first steps is to talk with the people involved with the process and to make a general list of the more common reasons why the product or service is not acceptable. Then take this list and begin to categorize the products or services that are not acceptable. This detection work must be done so

that in the near future proactive action can be taken to prevent the past big opportunities from occurring as frequently in the future.

4. Collect and record the data.

5. Maximize the ability to let the process "talk to you." There must be provisions to communicate who, what, when, where, and how. (There should be room for comments.)

Five types of check sheets are discussed in this section:

1. Defect type
2. Defect location
3. Check sheet for attribute measurements (go/no-go measurements)
4. Check sheet for variable measurements (multiple process streams)
5. Check sheet for truck line delivery performance

As we will shortly see, the main power of a check sheet is that the analysis of a check sheet after the operator has filled it in will usually give us a *Pareto distribution*. A Pareto distribution will point us in the direction where our efforts should be concentrated.

A check sheet, if properly created, completed, and finally analyzed, will tell us what type of defect to try to eliminate. It will tell where in the product the defect is consistently occurring or, in the case of a variable check sheet, where in the process we should make an adjustment, by how much, and in what direction.

When visual inspection or attribute measurements are taken or when variable measurements are made, or when the process is complex with multiple fixtures or cavities or numerous measurements on a single product, then check sheets are usually the answer, at least in the beginning.

Check sheets cannot take the place of variable control charts, nor can they give you a true prevention system.

9.2 Defect Check Sheet by Frequency and by Cost

Example 9.1 Figure 9.1 is an example of a weekly summary of a check sheet that is tracking the frequency of rejected crankshafts at final inspection. Multiple characteristics are being evaluated. Both variable and attribute product characteristics are being recorded by cause code. The cause codes are

1. Main bearing undersize
2. Main bearing taper
3. Main bearing out of round
4. Main bearing chatter
5. Main bearing not cleaned up

Weekly Crankshaft Reject Summary Report			
			Date _____
Reason for reject	Frequency	Cost	Total cost
M.B. U/S	3	$ 90.00	$ 270.00
M.B. taper	4	30.00	120.00
M.B. O/R	1	40.00	40.00
M.B. chatter	33	60.00	1980.00
M.B. NCU	12	190.00	2280.00
M.B. finish	2	12.00	24.00
Bad centers	0	25.00	00
P.B. U/S	5	90.00	450.00
P.B. taper	2	30.00	60.00
P.B. O/R	3	40.00	120.00
P.B. chatter	29	60.00	1740.00
P.B. NCU	2	190.00	380.00
P.B. finish	3	12.00	36.00

Figure 9.1 Weekly crankshaft rejection summary by frequency and by cost.

6. Main bearing finish
7. Bad centers on forging
8. Pin bearing undersize
9. Pin bearing taper
10. Pin bearing chatter
11. Pin bearing not cleaned up
12. Pin bearing finish

The five most frequent causes for rejection in descending order are

Main bearing chatter: 33

Pin bearing chatter: 29

Main bearing not cleaned up: 12

Pin bearing undersize: 5

Main bearing taper: 4

Visual analysis indicates that the big opportunities for improvement in lowering the rejection rate will require the reduction in frequency of "chatter" problems on both the main bearings and the pin bearings. The cost accounting department assigned a financial penalty for each of the cause codes. Some of the rejected products could be corrected by low-cost rework, while other reasons for rejection required that the finished crankshaft be scrapped. There is the possibility that during some of the rework operations the product might be scrapped. With a cost value given, we can multiply the frequency by the cost to end up with a Pareto analysis by cost.

The five cause codes for rejection in dollars are listed in descending order:

Main bearing not cleaned up	$2280.00
Main bearing chatter	$1980.00
Pin bearing chatter	$1740.00
Pin bearing undersized	$450.00
Pin bearing not cleaned up	$380.00

The "main bearing not cleaned up" condition was ranked third by frequency, but was ranked first in dollars. The main bearing chatter and pin bearing chatter were ranked 1 and 2, respectively, by frequency. These same characteristics were ranked 2 and 3, respectively, in dollars.

Whenever possible, the information from the check sheet should be assigned a dollar penalty, along with absolute frequency and by percentage. The analysis of the check sheets weighted by frequency and by dollars would change very little in this example.

Process improvement efforts should concentrate on reducing both the frequency and the dollars of loss due to

1. Chatter problems on both the main bearings and the pin bearings
2. Main bearing not cleaned up

Management should concentrate on the big opportunities for improvement.

The steps in *prevention* for a process currently producing a product or service are

1. Identify the *big problems*.
2. Find the *causes* of the big problems. Check sheets will help greatly.
3. Determine what to change in the process.
4. Measure the process performance after the change.
5. Fine-tune the process if needed.
6. Communicate the success.

The *main* purpose of a check sheet is to aid the user in gathering information so this information can be stratified (sorted) logically and then prioritized so that action can be taken on the process. Check sheets are a good way to acquaint people with thinking in a statistical way. Then later if needed, other statistical tools can be introduced on an as-needed basis.

Very little training is needed to use check sheets, and often check sheets by themselves identify the cause of the problem and help determine what action should be taken to solve the problem.

9.3 Defect Location Check Sheet

Example 9.2 A *defect location sheet* is shown in Fig. 9.2. The objective is to determine whether there is a nonrandom pattern in the location of the leak along a joint of a compressor that is bolted together. A cross section of the compressor

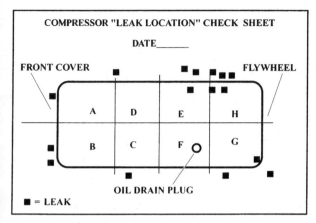

Figure 9.2 Defect (leak) location check sheet.

joint is drawn. The drawing is sectioned off into grids, A through H. When a leaky unit is repaired, the technician marks the location of the leak on the check sheet. As you can see, if we chart the frequency of leaks by location, we observe a nonrandom pattern. The opportunities for improvement in the frequency of leaks will require the discovery of those factors in the process which are causing a very high incidence of leaks in zones E and H.

Some possible causes might include

Flatness of mating surfaces

Surface finish of mating parts

Bolt torque

Bolt torque sequence

Gasket thickness

Gasket assembly method

Assembly tooling

Mating-part cleanliness

An in-depth investigation will be needed, and undoubtedly variable measurements will have to be taken, summarized, and analyzed. This check sheet is just the first of many steps that must be taken to improve this process.

> QUESTION: Is this not a situation where the design of experiment technique could be used to identify the factors that contribute to leaky compressors and the level of those factors that will minimize the chances of a leak happening?
>
> ANSWER: If the information obtained from measuring part flatness, finishes, torque, etc., does not identify where in the process action must be taken, then some form of designed experiment may be the next step. It makes more sense to first determine whether the assembly quality characteristics are

1. In control
2. On target
3. Capable

If the process satisfies these requirements, then design of experiment may be needed.

9.4 Attribute Check Sheet

Attribute-measuring instruments are used in many manufacturing operations. Some of the more common types of attribute gages are plug gages, snap gages, flush pin gages, paddle gages, and templet gages. Attribute gages should not be used on characteristics with tight tolerances or characteristics that are critical to product performance and durability. Often there is a need to document the measurements of numerous dimensions or characteristics that are checked with these types of go/no-go gages.

Figure 9.3 is a check sheet created for the documentation of many product characteristics measured with attribute (go/no-go) gages. A hydraulic pump housing is being die-cast. There are many dimensional checks that must be made periodically. The operator is instructed

GO-NO GO ATTRIBUTE CHECK SHEET				INSTRUCTIONS	
PART NUMBER ___ 7955___ PART NAME ___ housing ___				Measure the characteristics at the specified frequency if an out of specification product is found then perform 100% sorting back to the last good sample. Identify and quarantine any out of specification product.	
DATE_____ DEPARTMENT_____					
OPERATOR #_____ MACHINE #_____					
CHARACTERISTIC	FREQ.		PRODUCTION OPERATOR	INSP.
3.990–4.010 diameter	1/100	83401	OS OS OS OS OS OS OS OS OS OS OS OS (OK)(OK)(OK)(OK)(OK)(OK)(OK)(OK) OK OK (OK) OK US US US US US US US US US US US US	OS OS (OK) OK US US
6.470–6.530 length	1/100	60126	OS OS OS OS (OS) OS OS OS (OS) OS (OK)(OK)(OK)(OK) OK (OK)(OK)(OK) OK (OK) OK US US US US US US US US US US	(OS) OS OK OK US US
2.735–2.765 length	1/100	66762	OS OS OS OS OS OS OS OS OS OS (OK)(OK)(OK)(OK)(OK)(OK)(OK)(OK)(OK)(OK) US US US US US US US US US US	OS OS (OK) OK US US
3.235–3.265 shoulder length	1/100	66762	OS OS OS OS (OS) OS OS OS OS OS (OK)(OK)(OK) OK OK (OK)(OK) OK OK (OK) US US US (US) US US US (US)(US) US	OS OS (OK) OK US US
.370–.380 groove depth	1/100	66762	DP DP DP DP DP DP DP DP DP DP (OK)(OK)(OK)(OK)(OK)(OK)(OK)(OK)(OK)(OK) SH SH SH SH SH SH SH SH SH SH	DP DP (OK) OK SH SH
COMMENTS:				

Figure 9.3 Completed check sheet for attribute measurements (go/no-go measurements).

to measure the characteristics at the prescribed frequency and record the measurements. If the characteristic is out of specification, then the product produced since the last in-specification measurement must be 100 percent inspected for that characteristic. Any out-of-specification product must be identified and quarantined.

The ideal information on the check sheet would be that the operator found all the characteristics on all the products measured to be within the specification limits (OK).

1. If there is a repetitive pattern of a characteristic being oversized (OS), then this indicates that the tooling that controls the dimension for some reason is running off target on the high side.
2. A repetitive pattern of a characteristic found to be undersized (US) indicates that the tooling that controls that dimension is off target on the low side of the specification.
3. A repetitive pattern of a characteristic producing product out of specification on both sides—oversize and undersize—indicates that there is excessive variation.

In this die-casting example, the machine operator is very much at the mercy of the tooling provided. There are some tricks of the trade to compensate for tooling that does not produce in-specification product. But this will just make life more difficult on the factory floor for the manufacturing departments.

To evaluate a process, at least 1 or 2 weeks of completed check sheets is needed, before an analysis of the process can be made.

Check sheet analysis

The 3.990–4.010 diameter seems to be running okay. All the recordings indicate that the characteristic is within the specification. The 6.470–6.530 length appears to be running off target on the high side. Three oversized measurements are recorded on the check sheet. The 2.735–2.765 length seems to be running okay. All the measurements are within the specification limits. The 3.235–3.265 shoulder length appears to have excessive variation. One measurement is oversized, and two measurements are undersized. The groove depth seems to be running okay, and all the recorded measurements are within specification limits.

The next step in the evaluation of the process is to perform a more in-depth capability study of the 6.470–6.530 length and the 3.235–3.265 shoulder length. Arrangements must be made so that variable measurements can be taken. The first step is to determine whether these two characteristics are in a state of statistical control.

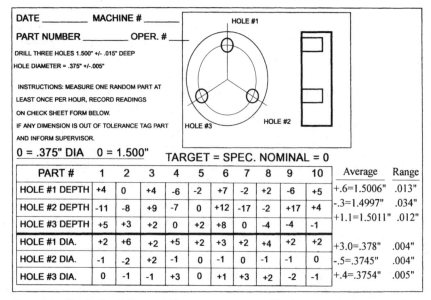

Figure 9.4 Check sheet for variable measurements with three process streams (completed form).

9.5 Variable Check Sheet

Figure 9.4 is an example of a *variable* check sheet.

An iron plate must have three holes drilled. Two characteristics are to be monitored:

1. Hole diameter: 0.375 ± 0.005 in
2. Hole depth: 1.50 ± 0.015 in

The part is being processed on a drill press with three separate spindles. There are three process streams in this case. There is a need to sort the measurements by stream. All parties (design, engineering, and manufacturing) agreed that the target is the middle of the specification. The manufacturing people will record the measurements to the nearest 0.001 in. The measurements are coded from zero. (Example hole depth of +4 = 0.379 in.)

A *good rule of thumb* is that if you have multiple process streams and numerous characteristics, a check sheet similar to the one in this example is the proper tool to use in the beginning of the data-gathering phase. This check sheet will point you to the specific quality characteristics of greatest concern.

Check sheet analysis

The ideal measurements on the completed check sheet would indicate that all quality characteristics were very closely centered on the target of zero and that there was small variation. Figure 9.4 is the completed check sheet.

A quick way to analyze the check sheet is to calculate the average and range for all six process streams.

Quality characteristic	Average	Range
Hole depth 1	+0.6 = 1.5006 in	13 = 0.013 in
Hole depth 2	−0.3 = 1.4997 in	34 = 0.034 in
Hole depth 3	+1.1 = 1.5011 in	12 = 0.012 in
Hole diameter 1	+3.0 = 0.378 in	4 = 0.004 in
Hole diameter 2	−0.5 = 0.3745 in	4 = 0.004 in
Hole diameter 3	+0.4 = 0.3754 in	5 = 0.005 in

Remember that this is a three-stream process (three independent spindles in the drill press) and that there are two quality characteristics of concern.

The variation (range) in hole 2 depth is very big. The observed range of the 10 measurements (0.034 in) is greater than the total tolerance of 0.030 in. Some type of maintenance or adjustment must be made to reduce the variation. Two products are out of specification, one on the low end and the other on the high end.

The diameter of hole 1 is off target on the high side by 0.003 in. There is one product out of specification on the high side, and there is one product measuring exactly at the high specification of +0.005. Investigation must be undertaken to determine the cause of this undesired condition so that corrective action can be initiated.

QUESTION: Does the information on the check sheet prove that the process is not capable?

ANSWER: The prerequisite for evaluating the capability of a process is first to determine whether the process is in a state of statistical control. Once the process shows a good state of control, then the capability of the process can be determined. The check sheet does give enough specific information so that action can be taken on the process now. Once the improvements have been made concerning the variation in hole 2 depth and correction for the diameter of hole 1, then control charts can be initiated.

We find ourselves in this predicament: There are two quality characteristics and three process streams. The effort of having six control charts is probably not warranted. Once the process has been improved so that all the streams for all the characteristics are well centered and exhibit about the same variation, possibly the process could be monitored by using only two control charts, one for hole depth and one for hole diameter.

There is enough information now to take action. Once that action has been taken and the improvement verified, then a determination can be made of how to monitor the process in the future.

The real power of check sheets is that if they have been properly constructed, completed, and analyzed, they give a Pareto distribution which tells us what is vitally important if we are to eliminate the big problems.

It is extremely important to stress that *someone must analyze the completed check sheets*. If we are going to train the operator or person performing the process to fill out the check sheet and give that person time during the workday to fill out the form, then the company must do the follow-up work and take action when needed. People want feedback about the analysis of the information. Do not think that the analysis must always tell us to fix a problem. The analysis might tell us that things are running well and that we should go find other opportunities for improvement. We should report on the analysis to the people regardless of whether the results are good or bad. If the results are bad, in some way we should let the operator know of the company's plan to improve the process.

9.6 Truckline Delivery Performance Example

A company is experiencing a high number of customer complaints for late delivery. The producing company is shipping the products with enough lead time for the truckline to deliver on time. Figure 9.5 is a run chart showing the percentage of on-time deliveries by month.

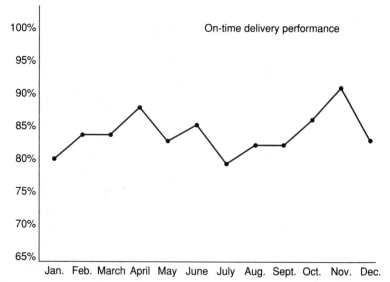

Figure 9.5 Run chart showing truckline delivery performance by month.

The traffic department wants to determine whether delivery performance is functioning in a state of statistical control. There should be 25 samples before statistical control can be determined with an attribute control chart. Regardless of whether the process is in control or out of control, obviously the level of performance is not acceptable. The traffic department realized that it could not wait 19 months before starting the corrective action efforts. Clearly the information on the run chart was not detailed enough to take action. It will be necessary to sort the data by truckline and by reason for late delivery.

Three trucklines are under contract for deliveries: Ray's Truckline, Bob's Truckline, and Joe's Truckline. Four reasons for late delivery were common to all three trucklines:

1. Mechanical breakdown
2. Driver change
3. Drop shipment
4. Schedule change

Figure 9.6 is a check sheet with the data stratified by truckline and cause code. The data should be sorted and summarized by

1. Total of late deliveries by truckline
2. Total of late deliveries by cause

	TRUCKLINE DELIVERY PERFORMANCE CHECK SHEET									
	MONTH	OCT	NOV	DEC	JAN	FEB	MAR			
	# OF SHIPMENTS	96	142	113	91	103	158			
	# DELIVERED LATE	35	51	26	33	21	47			
	% DELIVERED LATE	36%	36%	23%	36%	20%	29%			
Ray's Truckline	MECH. BREAKDOWN	2	4		3	1	2			
	DRIVER CHANGE	5	3	3		2	4			
	DROP-SHIP	5		2	1	2	1			
	TRUCK LINE SCHEDULE CHANGE	1	3		5		3			
Bob's Truckline	MECH. BREAKDOWN	6	11	5	4	6	5			
	DRIVER CHANGE	3	1		1	2	2			
	DROP-SHIP	1	2	1	2		8			
	TRUCK LINE SCHEDULE CHANGE	1	13	8	9	4	9			
Joe's Truckline	MECH. BREAKDOWN	1	2	3	4	2				
	DRIVER CHANGE	3	5	1	2	1	3			
	DROP-SHIP	3	2	3		1	8			
	TRUCK LINE SCHEDULE CHANGE	4	5		2		2			

Figure 9.6 Check sheet showing truckline delivery performance by cause code (three different trucklines).

3. Total of late deliveries sorted by truckline

The summary is as follows:
During the time period of October to March, a total of 703 shipments were made. During that time the trucklines were responsible for 213 late deliveries. And 30 percent of the shipments were delivered late to the customer.

Ray's truckline was responsible for 52 late deliveries during the 6-month period. Bob's truckline was responsible for 104 late deliveries. Joe's truckline was responsible for 57 late deliveries.

Mechanical breakdown was the cause for a total of 61 late deliveries. Driver change was the cause for a total of 41 late deliveries. Drop shipment was the cause for a total of 42 late deliveries. Schedule change was the cause for a total of 69 late deliveries.

Ray's truckline:

Mechanical breakdown was the cause for 12 late deliveries.

Driver change was the cause for a total of 17 late deliveries.

Drop shipment was the cause for a total of 11 late deliveries.

Schedule change was the cause for a total of 12 late deliveries.

Bob's truckline:

Mechanical breakdown was the cause for 37 late deliveries.

Driver change was the cause for a total of 9 late deliveries.

Drop shipment was the cause for a total of 14 late deliveries.

Schedule change was the cause for a total of 44 late deliveries.

Joe's truckline:

Mechanical breakdown was the cause for 12 late deliveries.

Driver change was the cause for a total of 15 late deliveries.

Drop shipment was the cause for a total of 17 late deliveries.

Schedule change was the cause for a total of 13 late deliveries.

Analysis

First, obviously there are grave problems with delivery performance regardless of how the data are stratified. The system for providing truck transportation as it exists now needs major changes. There is not one or two *major* opportunities for improvement. There are some areas to concentrate on that will have a small impact on improving the delivery performance.

The most obvious opportunity for improvement is that Bob's truckline has almost twice as many late deliveries as the other two trucklines. The policies, procedures, and process flows must be reviewed by all three trucklines, especially Bob's truckline. A member of the traffic department investigated further and determined that Bob's truckline was the prime truckline used by the company. Bob's truckline had more late deliveries because it delivered the great majority of shipments. So in actuality, Bob's truckline performance level was really no different from that of the other two trucklines.

The traffic department did not do a good job of planning how the data should be sorted and what level of detail would be required. This was a very costly, time-consuming, and expensive learning experience.

9.7 Summary

A check sheet is an excellent tool to use in the early stages of process improvement. A well-thought-out and -constructed check sheet will identify the major opportunities for process improvement. Whenever there are numerous process streams and/or multiple quality characteristics that need to be evaluated, a check sheet is the tool to use.

There is great emphasis on process capability today, and this emphasis leads to the use of Shewhart control charts and process capability studies. Process capability cannot be determined with the use of a check sheet. The analysis of the completed check sheet will usually lead to brainstorming sessions to determine what changes to make in the process and/or additional data that must be gathered.

Chapter 10

Scatter Plots

A scatter plot (regression analysis) is one of the seven tools of statistical problem solving. A scatter plot could be used to determine whether the mold temperature has an effect upon the flatness of an injection-molded part. In this case the independent variable is the process parameter of mold temperature, and the dependent variable is the quality characteristic of part flatness. If there is a strong cause-and-effect relationship, then the *target values for the independent variable can be set so that the objective of the quality characteristic being on target can be achieved.*

Many times we ask ourselves, How does changing one factor in the process affect the output of the process? That is,

- How does the cutting fluid affect tool life?
- How does employee training affect productivity?
- Is there a relationship between bolt torque and the percentage of power steering pumps that leak?
- Is there a cause-and-effect relationship between mold temperature and the brittleness of a molded plastic part?

Figure 10.1 shows numerous patterns that are commonly seen in a scatter plot. Situation A is described as a positive correlation—as X increases, Y also increases.

Situation B is referred to as *no correlation* between the two factors—no predications can be made. Situation C is a nonlinear relationship. Situation D is referred to as a *negative correlation*. As the X factor increases, the value for factor Y decreases. Situation E is another example of a nonlinear relationship. Situation F is usually described as an *exponential* relationship.

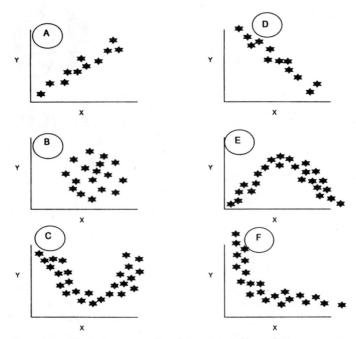

Figure 10.1 Most common scatter plot patterns of correlation.

10.1 Example

A manufacturer of starters for automotive engines is interested in finding out whether there is a relationship between the clearance between the pole shoes and the armature of the starter, and the amperage that the starter draws with the points open during testing at the factory. The objective of this exercise is to assemble starters that will "pull" 70 A.

To complete this problem, do the following:

1. Record the amperage and clearance for 10 units.
2. Plot the data on the scatter plot form.
3. Complete the work sheet portion.
4. Calculate the best-fit line.
5. Draw in the best-fit line.
6. Calculate the coefficient of determination.

7. Determine the clearance needed in the assembly to achieve the desire amperage.
8. Selectively assemble the starters to achieve the objective of 70 A.

Ten starters were serialized 1 through 10. The clearance between the pole shoes and the armature was measured and recorded. When the 10 starters were tested, the amperages were recorded. Both the clearance and the amperage could be traced back to the specific starters. The table shows the data from the 10 units.

Unit number	Clearance ($\times 10^{-3}$ in)	Amperage
1	3	88
2	7	70
3	5	76
4	6	80
5	9	64
6	4	84
7	6	74
8	3	91
9	5	75
10	8	80

The measurements for the two characteristics are now plotted on the scatter plot form in Fig. 10.2. By casual visual analysis, the 10 data points seem to follow a straight line. The least-squares method is used to determine the best-fit line.

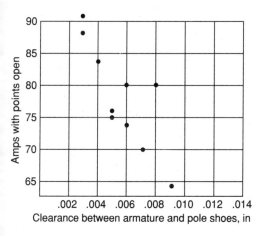

Figure 10.2 Scatter plot showing correlation between assembly clearance and amperage.

10.2 Calculating the Best-Fit Line

Starter number	X	Y	XY	X^2	Y^2
1	3	88	264	9	7744
2	7	70	490	49	4900
3	5	76	380	25	5776
4	6	80	480	36	6400
5	9	64	576	81	4096
6	4	84	336	16	7056
7	6	74	444	36	5476
8	3	91	273	9	8281
9	5	75	375	25	5625
10	8	80	640	64	6400
	$\sum x = 56$	$\sum y = 782$	$\sum (xy) = 4258$	$\sum (x^2) = 350$	$\sum (y^2) = 61{,}754$

Now that the table has been completed, values for b_1 and b_0 must be determined. The calculations are as follows:

$$b_1 = \frac{n(\sum XY) - (\sum X)(\sum Y)}{n(\sum X^2) - (\sum X)^2}$$

$$= \frac{10(4258) - 56(782)}{10(350) - 56^2} = -3.33$$

$$b_0 = \overline{Y} - b_1 \overline{X}$$

$$= 78.2 - (-3.33 \times 5.6)$$

$$= 78.2 - (-18.65) = 96.85$$

The next step is to choose two values for the clearance and calculate the corresponding amperage values. Once these two calculations have been made, the best-fit line can be drawn in. The two values chosen are clearances of 0.003 and 0.007 in. The calculations for units with clearances of 0.003 and 0.007 in are shown here:

$$y = b_0 + b_1 x$$

$$= 96.85 + (-3.33)(3) = 86.85$$

$$y = b_0 + b_1 x$$

$$= 96.85 + (-3.33)(7)$$

$$= 96.85 - 23.31 = 72.54$$

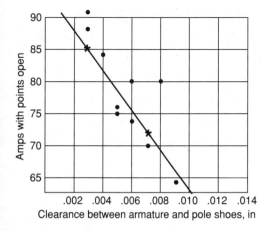

Figure 10.3 Scatter plot with best-fit line drawn in.

Based upon this best-fit line, we have determined that a unit with a clearance of 0.003 in will have an amperage value of 85.86 and that a unit with a 0.007-in clearance will have an amperage value of 72.54 A. These two points are plotted on the scatter plot, and the best-fit line is drawn in through these points. See Fig. 10.3.

10.3 Coefficient of Determination

The next step is to determine the extent to which the value of the dependent variable (clearance) affects the value of the independent variable (amperage). This is done by ascertaining the *coefficient of determination* R^2. The formula is

$$R^2 = b_1 \frac{n \sum (xy) - (\sum x)(\sum y)}{n \sum y^2 - (\sum y)^2}$$

$$= -3.33 \, \frac{10(4258) - 56(782)}{10(61{,}754) - 782^2}$$

$$= -3.33 \times 0.2014$$

$$= 0.671$$

The coefficient of determination is 0.671. In other words, 67.1 percent of the variation in amperage is defined by the prediction equation.

Recall that the objective is to learn what clearance in the starter (independent variable) assembly will translate to 70 A of power drawn (dependent variable). The procedure is as follows:

Locate 70 A along the Y axis.

Proceed directly to the right until the best-fit line is reached.

Move directly down until the X axis is reached.

By interpolation determine the clearance required.

Communicate this to the involved departments to ensure that the starters consistently draw 70 A of power.

> QUESTION: Is it an assumption that the other factors or components have little effect upon the amperage of the starter?
>
> ANSWER: In this example only two characteristics were evaluated, one dependent variable and one independent variable. Multiple regression can be performed on more than two variables if the situation warrants the effort. As the number of dependent and independent variables increases, probably a designed experiment would be a more appropriate tool.
>
> QUESTION: Instead of using regression analysis, would it be better to use control charts for the two characteristics? Then it would be possible to determine whether the two characteristics were
>
> 1. In control
> 2. Centered well on the target
> 3. Capable
>
> ANSWER: This is the question that needs to be answered in this situation: What is the relationship between the assembly clearance and performance of the product? One would think that if the assembly clearance were in control, on target, and easily meeting specifications, then the product would perform well. There is an assumption that the assembly clearance strongly influences product performance. The product design department probably used some form of scatter plot to determine the clearance that would translate to the desired performance (70 A).

A scatter plot is the appropriate tool to use at this stage. If it has been determined that there is no relationship, but the design department is sure that there should be, then additional investigation is needed.

10.4 Summary

A scatter plot should be used when there is a need to know how one independent factor in a process affects a quality characteristic. A scatter plot can also be used to determine whether there is a correlation between two measuring instruments. Many questions can be answered by just plotting the data and determining whether a correlation exists.

If a strong correlation does not exist, accurate predictions cannot be made. There are mathematical techniques for transforming data that do not correlate. Once this transformation has been made, then a prediction can be made. Finally, the reverse transformation of the data is made.

Whenever one asks, How does the changing of one factor in the process affect product performance? then the scatter plot is usually the statistical tool to use. But if the question is: How does the changing of eight different factors in the process affect the product performance? then a scatter plot is not the tool to use—design of experiment is.

Chapter 11

Design of Experiment

11.1 Introduction

This chapter explains a method of experiment design known as the *Taguchi method*. This technique can be used to optimize a manufacturing process or a product design. Figure 11.1 shows the principle of *leveraged quality*. In the ideal situation, the *voice of the customer* is obtained through some type of survey. The marketing, sales, and product development departments must make sure that the customers' wants are provided for in the product. The next phase will involve the product development department. At this phase the design of experiment

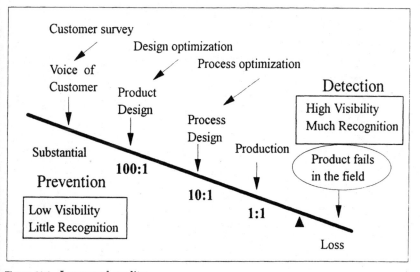

Figure 11.1 Leveraged quality.

technique should be used to optimize the design so that (1) the product satisfies the customer's requirement (target) and (2) there is little variation from one product to the next. During the product development phase, the design department should be very cost-conscious, using low-cost materials that do not have extremely tight tolerances. At first glance, high quality, high customer satisfaction, short product development time, low-cost materials, and relaxed specifications seem to conflict with one another. In reality, a combination should be the strategy used to ensure future competitiveness.

During the process design phase, design of experiment can be used to optimize the manufacturing process. This will reduce the likelihood of problems getting to the production area. During the production phase, statistical tools such as Pareto analysis, brainstorming, control charts, and check sheets are used.

The financial and quality leverage is gained by using customer survey techniques and design of experiment techniques at the early stages of product development. The worst place for a loss to occur is while the product is in the field. It is better, from a financial and a quality point of view, to detect the loss during production. The farther back on the quality level that the proper type of effort is expended, the greater the gains.

These are some of the common mistakes made during product development:

1. Take a "one factor at a time" approach to product development.
2. Design within specification limits, not to targets.
3. Work on only one product performance requirement at a time, instead of multiple performance requirements at once.

The product development engineer should wisely choose control factors and their different levels. One of the strategies promoted by Dr. Taguchi is to identify:

1. Control factors that affect targeting but not variation
2. Control factors that affect variation but not targeting
3. Control factors that affect both targeting and variation
4. Control factors that have no effect on either targeting or variation

Example 11.1 It is believed that the surface finish of a radius affects the machine efficiency. It is believed that the smoother the radius, the higher the efficiency. It is known that *manufacturing costs increase as smoother surfaces are specified.*

The current specification for radius surface finish is 63 arithmetic average (aa) maximum. Strategy 1 is the most common approach used by traditionally educated engineers.

Strategy	Control factor	Level 1	Level 2
Strategy 1	Radius surface finish	Current (63 aa max.)	Smoother (40 aa max.)
Strategy 2	Radius surface finish	Current (63 aa max.)	Rougher (100 aa max.)

Suppose that we use strategy 1. Our experimental results prove that surface finish has very little effect on machine efficiency. Have we gained any knowledge during our experiments that will give us an economic advantage? The answer is no. In this example, we did not choose a level for surface finish that would translate to a financial advantage.

With strategy 2, suppose that the experimental results again prove that radius surface finish has no impact on machine efficiency. Have we gained any knowledge that will give our company an economic advantage? Yes. This is the strategy suggested by Dr. Taguchi: If a control factor does not affect efficiency at either level, then choose the level that reduces costs or eases manufacturability.

And strategy 3 is an option: Experiment with the control factor at three levels rather than two.

Strategy	Control factor	Level 1	Level 2	Level 3
Strategy 3	Radius surface finish	Smoother (40 aa max.)	Current (63 aa max.)	Rougher (100 aa max.)

This third strategy (experimenting with the control factor at three levels) will give more information about the effect of radius surface finish on machine performance. We can determine whether radius surface finish has a linear or nonlinear effect upon machine efficiency.

A very costly approach still used by many companies today is to restrict tolerances more to ensure uniformity of product performance. In the ideal situation, the product development department should use designed experiments to help remove cost from the product with low-cost materials and components. Then the manufacturing engineering department should use the designed experiment technique to fine-tune the manufacturing processes.

Caution: Apply the Pareto principle when you choose where to spend the time, money, and effort of conducting the designed experiments. Ask yourself: What are the big opportunities in cost reduction, quality improvement, and reduced cycle time? Where should design of experiment be applied?

In Chap. 10, a scatter plot was used to evaluate the cause-and-effect relationship between one independent factor and one dependent factor. Usually there are many factors in a product design that affect product performance and reliability. Likewise, there are many factors in a manufacturing process that affect quality. An experienced engineer will be very knowledgeable about manufacturing processes.

Some form of multifactor designed experiments will identify which control factors have a strong impact on the quality characteristic of concern and what level is best for the control factors.

The following case study is an example of the Taguchi method of design of experiment. An experimental matrix (orthogonal array) is chosen depending on

1. The number of control factors
2. The number of levels for the control factors
3. Whether there is a need to measure the interaction between control factors

There are numerous orthogonal arrays to choose from. In this chapter we do not cover the unique intricacies of the different orthogonal arrays. In this chapter we limit ourselves to two case studies using design of experiment.

11.2 Case Study: Compression Strength

The objective in this case study is to increase the compression strength of molded briquette.

We know that the product is used as an alloying element in an iron foundry, and the briquettes are shipped in 50-lb bags. The customer's complaint is that a very high percentage of the product has crumbled into much smaller pieces or is even in a dustlike state. These crumbled briquettes cannot be used, and the employees are complaining about the dust.

The producer of the briquette acknowledged that it was having a problem with internal scrap due to the same type of defect.

A brainstorming session was held with the production people and other technical support groups. A *cause-and-effect diagram* was constructed. Originally more than 20 control factors were identified in the manufacturing process. With some reluctance, the group agreed to reduce the list to seven factors. The seven control factors were as follows:

A	Percentage of binder
B	Type of binder
C	Grain size
D	Percentage of moisture of feedstock
E	Oven temperature
F	Oven belt speed
G	Cooling rate

A decision was made to conduct a two-level experiment.

TABLE 11.1 Orthogonal Array with Test Results

Experiment no.	\multicolumn{7}{c}{Factor}	Results
	A	B	C	D	E	F	G	
1	1	1	1	1	1	1	1	315
2	1	1	1	2	2	2	2	420
3	1	2	2	1	1	2	2	195
4	1	2	2	2	2	1	1	370
5	2	1	2	1	2	1	2	290
6	2	1	2	2	1	2	1	565
7	2	2	1	1	2	2	1	480
8	2	2	1	2	1	1	2	540

	Factor	Level 1	Level 2
A	Percentage of binder	7%	11%
B	Type of binder	Current	New
C	Grain size	Small	Large
D	Percentage of moisture of feedstock	Wet	Dry
E	Oven temperature, °F	220	195
F	Oven belt speed	Fast	Slow
G	Cooling rate	Delayed	Normal

The orthogonal array in Table 11.1 tells us how to conduct the required experiments and the result of the experiments (in pounds of force required to fracture the sample):

$$L_8(2^7)$$

where 8 = number of experiments needed
 2 = number of levels for each factor
 7 = maximum number of factors in experiment

Each horizontal row in the array tells us how to vary the combination of control factors for each experiment. Notice that the first experiment tells us that all the control factors should be set at level 1: 7 percent binder, current type, small grain size, wet feedstock, 220°F oven temperature, fast belt speed, and delayed cooling time.

The average of the eight experiments $\bar{T}$ is 396.9 lb.

Next, we determine the average response for each factor at each of the two levels. The control factor was at level 1 for the first four experiments:

$$A_1 = \frac{315 + 420 + 195 + 370}{4} = \frac{1300}{4} = 325$$

$$A_2 = \frac{290 + 565 + 480 + 540}{4} = \frac{1875}{4} = 468.75$$

Recall that the quality characteristic (compression strength) is a larger-is-better type of characteristic. Factor A at level 2 is much better than factor A at level 1. The same calculations must be done for control factors B, C, D, E, F, and G.

$$B_1 = 397.5 \quad B_2 = 396.25$$
$$C_1 = 438.75 \quad C_2 = 355.0$$
$$D_1 = 320.0 \quad D_2 = 473.75$$
$$E_1 = 403.75 \quad E_2 = 390.0$$
$$F_1 = 378.75 \quad F_2 = 415.0$$
$$G_1 = 432.5 \quad G_2 = 361.25$$

Figure 11.2 shows the response plots for each factor. As the slope for a control factor gets steeper, that control factor has a stronger impact on the quality characteristic.

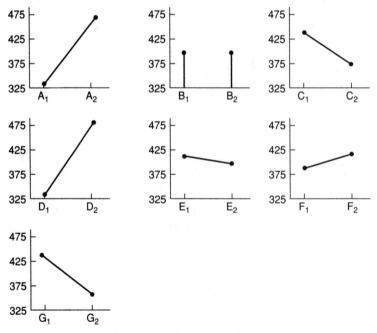

Figure 11.2 Response plot of compression strength for control factors.

The theoretical best combination of factors is

$$A_2, B_1, C_1, D_2, E_1, F_2, G_1$$

Notice that we are selecting the best level for each control factor. The theoretical best combination should give results that are greater than the best observed experimental results.

Prediction equation

Experience has proved that only about half the number of control factors manipulated during the experiment should be used in the prediction equation. If all the factors were used, the estimated average would be too optimistic and would not be reproducible during the confirming experiment. In this example there were seven control factors. The three control factors with the strongest effects on compression strength will be used to predict the estimated average $\hat{\mu}$.

The three control factors with the strongest effect are $\bar{D}_2, \bar{A}_1$, and $\bar{C}_1$. The formula for the prediction equation in this example is

$$\hat{\mu} = \bar{T} + \bar{D}_2 - \bar{T} + \bar{A}_1 - \bar{T} + \bar{C} - \bar{T}$$

$$= 396.9 + 473.75 - 396.9 + 468.75 - 396.9 + 438.75 - 396.9$$

$$= 587.45$$

The prediction equation tells us to expect an average compression strength of 587 lb. This is only a theoretical estimate based upon the results of the eight experiments. A confirming experiment must be run to verify that the results are as expected.

This case study was very simple and straightforward. Very seldom will things be as simple as this. Each experiment was conducted only once, control factor interactions were not measured, and no method of data transformation was chosen. In this example a control factor is assigned to each column in the orthogonal array matrix. This is an example of a *saturated experiment*. The orthogonal array is completely filled up with control factors.

11.3 Data Transformation and Analysis

Experience has shown that by transforming raw measurements into a more additive form we can accurately predict the optimum combination of factor levels more accurately than if the raw measurements were used directly. Remember, one objective is to identify a product design or a manufacturing process whose factors possess *additivity*. If we do not transform raw data properly, the predicted additivity will not agree with actual real-world observations.

There are many methods of transforming raw data. Some of the more common are the

- Log transformation method
- Omega conversion method
- Accumulation analysis method

In most situations, raw data do not exhibit additivity. Dr. Taguchi's years of experience have proved that data transformation usually provides characteristic values that better suit the experimental objective. The log transformation method will be shown on the following pages for a larger-is-better characteristic only.

Time has shown variable-type quality characteristics to be the most successful. On the other hand, when the quality characteristics under evaluation are attribute, the improvements are usually not as great as when the characteristics are variable. This does not mean that we cannot use the design-of-experiment tool when the quality characteristic is attribute. We should have lengthy discussions in brainstorming groups to try to come up with a substitute quality characteristic that is variable. This does influence the performance of the observed characteristic that is attribute.

Signal-to-noise ratio

Each of these characteristics has a different set of formulas used to calculate the signal-to-noise ratio. In all cases, the larger the signal-to-noise number, the higher the level of quality will be. This also translates to higher customer satisfaction and reduced variation around the target.

The advantages of using the signal-to-noise ratio, compared to regular analysis, are as follows:

- Signal-to-noise ratio calculations consider both the mean and the variation of the process.
- The signal-to-noise ratio is a form of data transformation that gives us a measure of performance of the effect of noise factors upon the process.
- The signal-to-noise ratio has a direct relationship to cost (Taguchi loss function).

The units of the signal-to-noise ratio are in decibels (dB), and the signal-to-noise ratio is denoted by η (Greek eta). This method transforms the raw observed measurements into logarithmically adjusted values.

Larger is better

The goal in this situation is to increase the mean and reduce variation.

$$\eta = -10 \log \left[\frac{1}{n} \sum \left(\frac{1}{y_1^2} + \frac{1}{y_2^2} + \cdots + \frac{1}{y_n^2} \right) \right]$$

We want to increase the number of cycles an assembly can be actuated before failure. Three assemblies were tested to failure. The life of the units in cycles was: 916, 853, 1145.

$$\eta = -10 \log \left[\frac{1}{n} \sum \left(\frac{1}{y_1^2} + \frac{1}{y_2^2} + \cdots + \frac{1}{y_n^2} \right) \right]$$

$$= -10 \log \left[\frac{1}{3} \sum \left(\frac{1}{916^2} + \frac{1}{853^2} + \frac{1}{1145^2} \right) \right]$$

$$= -10 \log \left[\frac{1}{3} \sum \left(\frac{1}{839{,}056} + \frac{1}{727{,}609} + \frac{1}{1{,}311{,}025} \right) \right]$$

$$= -10 \log \left[\frac{1}{3} \sum (0.000001191 + 0.000001374 + 0.000000762) \right]$$

$$= -10 \log \frac{1}{3} (0.000003327)$$

$$= -10 \log 0.000001109$$

$$= -10 \, (-5.955)$$

$$= 59.55$$

11.4 Case Study: Burst Strength

A company is having problems with low burst strength of an injection-molded plastic canister. Eleven processing factors were identified to be used during the experiments. A decision was made to use a two-level factor experiment. The factors and levels are listed here:

	Factor	Level 1	Level 2
A	Mold temperature	70	95
B	Melt temperature	480	440
C	Probe temperature	450	475
D	Injection pressure 1	65%	75%
E	Injection pressure 2	54%	60%
F	Injection speed	70%	55%
G	Injection time	5 s	4 s
H	Hold time	10 s	15 s
I	Cooling time	15 s	10 s
J	Booster pressure	Current	New
K	Flow rate	4.4	6.0

The product is being produced by using a two-cavity die. Variation from cavity to cavity is considered a noise factor. The second noise factor is the variation in material from supplier to supplier.

There will be a total of four products destructively tested for each of the 12 experiments. Two products from each material supplier will be tested. Both cavities will be evaluated. An L_{12} orthogonal array is appropriate in this situation. The layout of the experiment follows with the results from the four tests for each experiment.

Section A of Table 11.2 is the orthogonal array showing the layout of the 12 experiments and the combination of factor-level settings for the experiments.

In section B of Table 11.2, the measurements of the 48 burst tests (four products for each of the 12 experiments) are boxed in.

The first column is from cavity 1, supplier A.

The second column is cavity 2, supplier A.

The third column is cavity 1, material B.

The last column is the measurement from cavity 2, material B.

In section C of Table 11.2 the totals, averages, and signal-to-noise ratios for each experiment are calculated.

In section D of Table 11.2, we begin the process of determining the average main effect for each factor level. Factor A is at level 1 for the first six experiments, so we calculate the average signal-to-noise ratio value when A is at level 1. This calculation must be repeated for all 11 control factors. Recall that the larger the signal-to-noise ratio, the higher the level of quality.

Note! The best (highest) observed signal-to-noise ratio value was experiment 7 with a value of 51.89 dB. The grand average of all the experiments is 50.52 dB. Remember, this is a larger-is-better (LIB) type of characteristic.

Now we begin to determine the optimum or near-optimum combi-

TABLE 11.2 Layout of Experiments Including Measurements, Totals, Averages, and Signal-to-Noise Ratios

A = L12 ARRAY / BURST STRENGTH — CONTROL FACTOR
B = NOISE FACTORS CAVITY # /MATERIAL
C = TOTAL / AVERAGE
D = LARGER IS BETTER S/N RATIO

EXP #	A	B	C	D	E	F	G	H	I	J	K	1-A	2-A	1-B	2-B	TOTAL	AVERAGE	S/N RATIO
1	1	1	1	1	1	1	1	1	1	1	1	345	340	370	400	1455	363.75	51.16
2	1	1	1	1	1	2	2	2	2	2	2	360	315	335	355	1365	341.25	50.62
3	1	1	2	2	2	1	1	1	2	2	2	320	320	335	280	1255	313.75	49.87
4	1	2	1	2	2	1	2	2	1	1	2	375	310	330	265	1280	320	49.9
5	1	2	2	1	2	2	1	2	1	2	1	305	320	375	350	1350	337.5	50.48
6	1	2	2	2	1	2	2	1	2	1	1	305	350	370	295	1320	330	50.26
7	2	1	2	2	1	1	2	2	1	2	1	405	375	410	385	1575	393.75	51.89
8	2	1	2	1	2	2	2	1	1	1	2	245	305	260	365	1175	293.75	49.06
9	2	1	1	2	2	1	2	2	1	1	1	280	405	310	280	1275	318.75	49.79
10	2	2	2	1	1	1	1	2	2	1	2	385	400	330	365	1480	370	51.3
11	2	2	1	2	1	2	1	1	2	2	1	340	405	385	410	1540	385	51.64
12	2	2	1	1	2	1	2	1	2	2	1	330	360	290	340	1320	330	50.29

GRAND AVERAGE= 343.46 PSI   50.52 db

nation of process parameter settings. We are concerned with the average signal-to-noise ratio for each factor at both levels.

Table 11.3 is the completed response table showing the average signal-to-noise ratio values for all 11 factors at the two levels.

Hint: The total average signal-to-noise ratio for the two levels in this example will be 101.04 for all the factors. So once we have determined the average for a factor at level 1, we can subtract that value from 101.04. The remainder will be the average effect for that factor at level 2. This shortcut works only when we are evaluating two-level factors.

The theoretical best combination of process settings is boxed in Table 11.3:

$$A_2, B_2, C_1, D_2, E_1, F_1, G_1, H_2, I_1, J_2, K_1$$

We can predict the estimated results of the process if we set the process at the best combination. This is where the additivity effect of taking advantage of combinations of factors is seen. Of 11 factors, we will pick those five factors that have the largest significant impact on burst strength:

$$E_1, F_1, G_1, I_1, J_2$$

TABLE 11.3 Response Table of Compression Strength

	SIGNAL TO NOISE AVERAGE RESPONSE		TOTAL
	A1=50.38	A2=50.66	101.04
	B1=50.40	B2=50.64	101.04
	C1=50.57	C2=50.48	101.04
	D1=50.49	D2=50.56	101.04
*	E1=51.14	E2=49.90	101.04
*	F1=50.73	F2=50.31	101.04
*	G1=50.71	G2=50.36	101.04
	H1=50.38	* H2=50.67	101.04
*	I1=50.69	I2=50.35	101.04
	J1=50.24	* J2=50.79	101.04
	K1=50.64	K2=50.40	101.04

The average signal-to-noise ratio of all experiments $\overline{T}$ is 50.52 dB.

Estimated average = $\hat{\mu}$

$$\hat{\mu} = \overline{T} + E_1 - \overline{T} + J_2 - \overline{T} + I_1 - \overline{T} + F_1 - \overline{T} + G_1 - \overline{T}$$

$$= 50.52 + 51.14 - 50.52 + 50.79 - 50.52 + 50.69$$
$$- 50.52 + 50.73 - 50.52 + 50.71 - 50.52$$

$$= 51.98 \text{ dB}$$

Using the reverse transformation method, we predict that the estimated average burst strength is 397 lb/in^2.

We should consider internal production cost and productivity when assigning the levels for the factors that do not significantly affect burst strength.

11.5 Summary

In today's world of intricate product designs and complicated manufacturing processes, the competitive company must use some form of designed experiments to quickly and economically determine the optimum combination of design parameters or manufacturing process parameter settings.

Changing one factor at a time is very time-consuming, has poor reproducibility, and does not take advantage of the additivity effect of control factors.

In the past 10 or 15 years, the term *robust design* has come into prominence. This approach is strongly associated with the philosophies and techniques developed by Dr. Genichi Taguchi. Some things to remember concerning robust design are as follows:

- Greater importance is placed on reducing variation with Taguchi's methods than with the classical methods of design of experiment. A financial penalty can be calculated based upon the targeting of a process and the amount of variation in the process.

- Orthogonal array matrices allow for the evaluation of many control factors while requiring a minimum of experiments. Techniques are available to evaluate the effect of interactions of control factors. Techniques have been developed to minimize the effect of control factor interaction while conducting the experiments. These techniques can be applied in the real world.

- A factor which cannot be controlled or is too expensive to control but has an impact upon the quality objective should be included in the designed experiment as a noise factor. The objective when one is using robust design is to identify the best combination of control factors that affects the quality characteristic and is also insensitive to noise factors (robust).

- If a control factor has little effect on the quality characteristic, then a decision of what level to choose for that control factor should be based upon cost savings and productivity improvements. It is common for the main objective in designed experiments to be cost reductions, improvements in productivity, and higher levels of quality.

Index

Actionable tasks, 28, 30, 33

Brainstorming, 27–29
  guidelines for, 29

Capability, 7, 8, 63, 97
  calculations, 107, 148, 151–153, 170, 179
  capability ratio, 150, 151
  cost considerations, 146
  C_p index, 143
  C_{pk} index, 147
  goals, 146
  graphically explained, 144
  relationship to customer satisfaction, 145
  using normal probability paper to determine, 127, 155, 156
Cause and effect, 28, 29
Cause code menu, 24
Central limit theorem, 114, 115, 177
Checksheets, 213–225
  attribute type, 218, 219
  defect location type, 216–218
  measuring on-time delivery, 222–225
  steps in constructing, 213, 214
  variable type, 220, 221
Contingency planning, 41–43
Continuous improvement:
  graphically explained, 53, 54
  methods of summarizing, 165
Control charts:
  for attribute characteristics, 191–211
  c chart, 205–211
  common types of, 192
  np chart, 204
  p chart, 194–199
    sample size considerations, 193
    variable sample size, 199–203
Control charts for variable characteristics, 47
  calculations for control limits, 105
  case study, 103–107
  common types of, 95, 96

Control charts for variable characteristics (*Cont.*):
  determining capability from, 107
  factors used when calculating limits, 142
  foundry example, 171–178
  goals when using, 102
  individual and moving range, 124–127
    calculation of limits, 126
  for low volume application, 137–140
  median and range, 167–170
  modified limits, 131–134
  moving average and moving range, 128–131
    advantages of, 129
  multistream chart, 134–137
  number of measurements needed, 122, 123
  power of, 102
  preplanning checklist, 102
  recalculation of control limits, 110–113
  recording of process changes, 159
  sampling for continuous processes, 123, 124
  sampling guidelines, 114
  sampling schemes, 115–121
    instant time sampling, 116
    over time sampling, 116
  sensitivity, 125
  using menu system to record process changes, 160
  visual pattern analysis, 125, 127, 161–165
Cross-functional teams, 7
Competitiveness factors affecting, 2
Cost of poor quality, 9–11
Customer satisfaction, 14

Data:
  actionable, 23
  methods of evaluation, 46
  organized, 45
  stratified, 22
Deming, Dr. W. Edwards, 12, 27, 33, 159

Design of experiment, 7, 12, 235–247
  data transformation, 241–243
  noise factor, 12, 244
  orthogonal array, 245
  prediction equation, 246
  selecting factor levels, 236, 237
  signal-to-noise ratio, 242, 243
Detection, 51
Detection management, 8, 17
Distribution:
  bimodal, 64, 65
  normal, area under curve, 51, 53
  skewed, 46, 62
  shape of attribute characteristics, 194

Focused thinking, 27

Gage correlation, 85
Gage repeatability and reproducibility (gage R&R), 83
Gantt chart, 41, 42
Goals, long range, 4

Histogram, 47
  considerations when constructing, 55
  determining number of cells, 56, 57
  skewed distributions, 62
How-why diagram, 28, 33–41

Ishikawa diagram, 28, 29

Juran, J. M., 21

Local workforce, 4

Matrix model, 28, 29–33
Measurements, variable, 45
Measurement system analysis, importance of, 81
Measurement system evaluation:
  for attribute characteristics, 88–92
  preparing for, 83
Measures of central tendency, 48
  median, 48
Measures of dispersion, 49–54
  range, 49

Measures of dispersion (*Cont.*):
  standard deviation, 49–54
    formula, 50, 59, 143
Multivary charting, 180–183
  sample size requirements, 181
Mil-std 414, 122, 123

Non-value-added processes, 11
Normal probability paper, 66, 67

Pareto, Vilfredo, 21
Pareto analysis, 21, 48, 214, 215
  distribution, 214
  principle of, 19
  of work order forms, 25–27
Planning long range, 4, 5
Precontrol, 185–188
  determining warning limits, 186
Prevention, 51
Prevention management, 8
Proactive management, 5–8, 18, 24
Process control, 18
  graphically described, 13, 14
Process improvement, 12
  implementation of, 14
  top management responsibility, 14, 15

Quality characteristics, types of, 70
Quality function deployment, 6, 14
Quality leveraged, 235

Reactive management, 8, 18
Regression analysis (*see* Scatter plot)
Run chart, 35, 47, 60, 61

Scatter plot, 227–233
  calculations for best fit line, 230
Shewhart, Dr. Walter A., 45, 60, 95
Statistical control, 7, 8, 97
Statistical errors, 82, 118, 123, 124
Statistical problem solving:
  defined, 1
  list of tools, 1
Statistical process control, 12
Strategic business planning, 7
Strategic objectives, 10
Strategy, long range, 18

Taguchi Dr. Genichi, 235
  formula, 70–76, 108, 109, 173
  loss function, 17, 67–76
Target value, 9, 14–18, 49
Total Quality Management, elements of, 2–4, 18

Variation:
  common cause, 47, 97, 101, 102, 113, 140

Variation (*Cont.*):
  piece-to-piece, 119, 180
  within product, 48
  special cause, 47, 97, 101, 102, 126, 140
  stream-to-stream, 181
  time-to-time, 119, 181
Voice of the customer, 16

Z formula, 52, 147
Z table, 79

ABOUT THE AUTHOR

Thomas J. Kazmierski is president of Total Quality Management Institute, and has more than twenty years of experience in quality assurance and manufacturing. He has taught more than 350 seminars and consulted with companies in the United States, Canada, Mexico, Israel, and Europe. Previously, Mr. Kazmierski served as quality engineering manager for Navistar's engine plant in Illinois, and worked for both General Motors and Sperry Rand. He also taught in the department of technology at Southern Illinois University.